Can a Bee Sting a Bee?

And Other Big Questions from Little People

CAN A BEE STING A BEE?

And

Other **BIG** Questions

from *Little* People . . .

Compiled by
Gemma Elwin Harris

An Imprint of HarperCollinsPublishers

HarperCollins books may be purchased for educational, business, or sales promotional use. For information please e-mail the Special Markets Department at SPsales@harpercollins.com.

First published in Great Britain in 2012 by Faber and Faber Limited.

A hardcover edition of this book was published in 2012 by Ecco, an imprint of HarperCollins Publishers.

FIRST ECCO PAPERBACK EDITION PUBLISHED 2014.

Library of Congress Cataloging-in-Publication Data has been applied for.

ISBN 978-0-06-222323-4

16 17 18 OV/RRD 10 9 8 7 6 5 4 3 2

The pursuit of truth and beauty is a sphere of activity in which we are permitted to remain children all our lives.

Albert Einstein

To Evie, Rosie, Eliza and Seth

CONTENTS

EDITOR'S NOTE

My son is two years old and already the questions have started. Recently he pointed at the moon as we hurried home from nursery and asked, 'What dat?' For now, 'That's the moon' will do as answer, but I know it won't be long before I'm struggling to explain what the moon is made of, how far away it is, and whether a goldfish could survive there.

The questions children ask are often baffling. Chances are, if you ever knew the answer – or even part of the answer – you've probably forgotten it or can only remember a half-baked version of the truth. Imagine if you could turn to a well-known expert at this point and get them to answer for you, in language simple enough for a child to understand. This was the idea behind **BIG QUESTIONS**.

With the help of ten elementary schools, we asked thousands of kids between the ages of four and twelve to send us the questions they most wanted answered. The results were fascinating and funny. There were cute and quirky questions, like 'Why is space so sparkly?', 'Who had the first pet?' and 'Can a bee sting a bee?' Others were fiendishly difficult: 'How is electricity made?' or 'Where do oceans come from?' And a few shot straight to the heart of a deep philosophical conundrum: 'Why do we have wars?', 'How do we fall in love?' and 'Where does good come from?'

Among their hand-written replies we found a lot of questions involving bodily functions. 'Why is wee yellow?' seemed to be a recurring concern. The mysteries of space clearly obsessed many children, and it's no surprise that animals – chickens, cows and monkeys – popped up frequently. There was even one question, of great genius, that encapsulated all the above – a perfect storm of cows, bowels and space travel: 'If a cow didn't fart for a year and then did one big fart would it fly into space?'

What would world experts say, when faced with these questions? The response from our panel has been staggering and heart-warming. However busy, they've carved out time to co-write this book in order to benefit the NSPCC – the UK's leading child protection charity.

Bear Grylls took the trouble to explain the nutritious benefits of eating a worm. Jessica Ennis emailed a mantra for aspiring Olympians, just two months before the 2012 Games. Derren Brown set his impressive grey matter to work on 'Is the human brain the most powerful thing on earth?' While Philippa Gregory put her latest novel on hold to shed light on why Guy Fawkes was 'so naughty'. No question was too bizarre. The historian Bettany Hughes barely blinked when we asked her 'Did Alexander the Great like frogs?'

This book doesn't claim to offer the only answers to these

questions. It's an anthology of voices, a personal response from each expert to a child's idiosyncratic question. We hope you enjoy reading them with your family and take something from them – including the mental image of a cow soaring into the stratosphere powered by its own methane. (Thanks to the science writer Mary Roach and her friend Ray, a real life rocket scientist, for running the maths on that one.)

When my son asked his question about the moon that evening, I was busy making a mental list of what we had in the fridge for dinner. Lying back in his buggy, he was taking in the beauty of the night sky. There above, he saw a pale and ghostly globe shining in the darkness for the very first time. His question 'What dat?' demanded I see that full moon too. So we stopped and stared, and how strange and new it seemed to us both.

Gemma Elwin Harris

A **BIG** THANK YOU

I cannot thank enough the brilliant and extremely busy scientists, historians, philosophers, psychologists, naturalists, explorers, artists, musicians, authors, archaeologists and paleontologists, sportsmen and women who took time to answer a child's question for this book. As well as the much-loved comedians who wrote one-liners for our Out-Takes chapter. It would take too long to thank you each individually here but I'd just like to say how much your kindness has been appreciated by the NSPCC.

Without the enthusiastic participation of ten schools, we'd have had no children's questions to ask. So a special thank you to the staff and pupils at: Corstorphine Primary, and Mary Erskine and Stewart's Melville Junior School, Edinburgh; Cleobury Mortimer Primary School, Shropshire; Woodland Grange Primary, Leicester; Furzedown Primary School, Tooting; Raysfield Infants' School, Chipping Sodbury; The Mulberry Primary School, Tottenham; Shottermill Junior School, Haslemere; Boxgrove Primary School, Guildford; and Grange Primary, Newham. In particular to Gillian Lyon the deputy head at Mary Erskine and Stewart's Melville, Caroline Gorham, and Ed Flanagan and Kirk Hayles at Woodland Grange Primary for their **BIG QUESTIONS** assembly.

To friends whose curious offspring, nieces and nephews got the first questions rolling in: The Scott clan, the Wrays, the Flemings, Lucinda Greig and extended family, Melonie Ryan, Wendy and Alfie Carter, Cat Dean and brood, Nicole Martin, Ben Crewe and Ruby, Esther and Hannah Davis.

For ideas, advice and introductions: Yana Peel of Outset UK, Joe Galliano, Simon Prosser, Jamie Byng, Marcus Chown, Duncan Copp, Chris Riley, Richard Holloway, Justin Pollard, Roger Highfield, Chris Stringer, and Giles Morgan at HSBC sponsorship. Not forgetting dear friends Gus Brown, Sally Howard, Amy Flanagan, Ngayu Thairu, Chris Hale, Catherine and Ralph Cator, Bex and Adam Balon, and my sisters Sophie and Lucinda whose encouragement, contacts and science-brains I'd have been lost without.

I'm grateful to those patient agents who went above and beyond: Jo Sarsby, Nelle Andrew, Sue Rider and Sophie Kingston-Smith, Stephen at Michael Vine Associates, Catherine Clarke, Caroline Dawnay, Hannah Chambers and Vivienne Clore.

A heartfelt thank you to my own agent Gordon Wise at Curtis Brown and to my editor Hannah Griffiths for embracing the project straight off and devoting careful thought and creativity to it ever since. To the team at Faber & Faber: Lucie Ewin, Donna Payne, Sarah Christie; and our illustrator Andy Smith.

Also to Kristine Dahl at ICM and Hilary Redmon at Ecco, HarperCollins, for their commitment and vote of confidence from across the pond. Everyone involved in **BIG QUESTIONS** is massively excited to be taking it to the US.

Which brings me to the sparky teams at the NSPCC. My respect and admiration goes out to you and I hope the proceeds from this book go some way to supporting the vital work you do every day. Charly Meehan, Viola Carney, Stefan Souppouris, Helen Carpenter, Lucie Sitch, Sarah Dade, Dan Brett-Schneider and the Fundraising Communications Team – you've been fantastic to work with.

Finally, love and thanks to my gorgeous husband, Nick. It would take a book to list the reasons why.

ARE THERE ANY UNDISCOVERED ANIMALS?

Sir David Attenborough

naturalist

Yes. Hundreds. Probably thousands. Exactly how many no one can say – because they have not yet been discovered.

If you spent a day in a tropical rainforest, swishing a butterfly net around through the undergrowth or the leaves high in the tree canopy, you would certainly collect hundreds of insects. Many of them would be beetles. Would any of them be unknown to science? You would have to ask a beetle scientist. Many he would recognise straight away. But there might be a few that would puzzle him.

Would they be new species? It might take him a long time in a museum, examining them and comparing them with others in the museum or pictured in books about beetles, to be quite certain that he had a new species. But there would probably be one. In fact, I suspect it might be harder to find a beetle scientist who would be able to do this very difficult work than to find an unknown beetle.

Unknown big animals are certainly much rarer. Your best

chance of finding one would be to go to the least explored part of our planet – the very deep sea. You can only go down there in special deep-sea submarines. They have to be extremely strong to withstand the huge pressure of the water. And of course it is pitch black down there, so you would have to have powerful lights to do your searching.

You might get a glimpse of something in their beams, but unless you could catch it and examine it in detail, you could not be certain that it was a new species. And catching animals down there is a very difficult job, needing very specialised equipment. But I am sure there are still monsters down there that no one has ever seen before.

IS IT OK TO EAT A WORM?

Bear Grylls

explorer and survival expert

Well, here's the thing . . . If your life depends on it, then you bet it is OK to eat a worm. But you don't want to be doing it every day, trust me. And if you do eat one, you've got to be careful because worms can have some bad stuff in their tums (as they wiggle around all day underground!) So it's best to cook them up. I find if you boil them up with some pine needles over a fire, it makes them taste a little bit better.

I will never forget the first worm I ate. I was standing there, incredulous, watching this soldier suck a long, juicy worm up between his teeth and munch it down raw. I was almost sick. When it was my turn, I think I nearly *was* sick.

But guess what? If you do it enough and you are hungry enough, then it gets easier. And there is the real secret of life and survival: if your spirit is strong enough, you will find a way to do the impossible. That's the lesson of the worm. Oh, and remember: keep smiling even when it's raining. That's the second-most important lesson. So get out there and explore!

WHAT ARE ATOMS?

Marcus Chown

author of books about space and the universe

Atoms are the building blocks out of which everything is made: you, me, trees, even the air we breathe. You cannot see atoms because they are very small. It would take ten million laid end to end to stretch across the dot of the exclamation mark at the end of this sentence!

But if you could see an atom, you would notice something very odd indeed. They are not made of much at all. In fact, they are pretty much all empty space.

At the centre of an atom is a tiny speck of matter called a nucleus. Circling it, like planets around the Sun, are even tinier specks of matter known as electrons. But in between the nucleus and the electrons is a lot of empty space. It means that you and I – since we are made of atoms – are mostly empty space.

In fact, there is so much empty space inside atoms that if you were to squeeze out all the empty space from all the atoms in all the people in the world, they would fit in the volume of a sugar cube. Imagine. The whole human race squeezed down to the size of a sugar cube. Mind you, it would be a very, very heavy sugar cube!

One more thing about atoms. They come in ninety-two different types (plus a few kinds that do not exist in nature but that scientists have made). And, just like if you put together different combinations of Lego bricks, you can make a house or a dog or a boat, atoms go together in different combinations to make a rose or a tree or a newborn baby. All of us are combinations of atoms. We are all different from each other because we are all different combinations of atoms.

WHY ARE THE GROWN-UPS IN CHARGE?

Miranda Hart

comedian, writer and actress

I have to say, I do sometimes wonder myself. Maybe you have asked this question because you have seen grown-ups doing things you don't understand or telling you to do things that don't seem right or fair. I'm sure you think you would be much happier if you didn't have to do what they tell you. And sometimes, although I am supposedly one of the grown-ups, when someone older than me, or in a superior job to me, tells me what to do, I can get very angry and think they are wrong.

But here is the thing. We have to trust that people older than us have so much experience and wisdom of life that they are making the right decisions, with our safety and interests at heart because they love us. It may not always feel that way at the time and sometimes grown-ups do get it wrong. If you strongly disagree then you must calmly tell them without getting cross and see what they say. But basically, as people get older, their life experiences mean they are cleverer and know best and that's why they have to be in charge. One day when you are a grown-up you will realise exactly what I mean.

I do, however, want to share one little secret with you. I think adults can go wrong because they forget what it's like to be a child. So you can remind the grown-ups of three key things:

Firstly, that it's important to take time out to play with you. Because sometimes they can work too hard.

Secondly, remind them to stop worrying what others think of them, just be themselves and boldly claim what their dreams are. It's very silly not to follow your dreams, don't you think?

And lastly, you can teach them to take each day at a time, milk every last bit of fun from it, and not worry about tomorrow. Because grown-ups forget to be free and joyful in the moment and you are brilliant at that.

WHY IS BLOOD RED, NOT BLUE?

Dr Christian Jessen

medical doctor and broadcaster

You may have heard that kings and queens have blue blood. That would be funny, but I'm afraid it isn't true. Nobody has blue blood. It's always red.

I know that if you look closely at the veins in your arms it seems like they contain blue blood. But this is just because your veins are very close to the surface of your skin, and the skin only lets certain colours of light through it – so the blood inside appears blue from the outside. Inside your veins, though, it's still red.

What gives your blood its red colour? The colour comes from a very important chemical in blood called haemoglobin, which carries the oxygen from your lungs all around your body, giving you lots of energy to move. Although it is *never* blue, haemoglobin can change colour a bit. When there is lots of oxygen in your body, your haemoglobin will make your blood a nice bright red colour. If you are running and playing, your body uses up more oxygen and your blood becomes a much darker red, and is quickly pumped back to your lungs to get some more oxygen.

Some animals *do* have blue blood, however. Do you know which ones? Octopuses, squid, lobsters, cuttlefish and horseshoe crabs all have blue blood!

HOW ARE DREAMS MADE?

Alain de Botton

philosopher

Most of the time, you feel in charge of your own mind. You want to play with some Lego? Your brain is there to make it happen. You fancy reading a book? You can put the letters together and watch characters emerge in your imagination.

But at night, strange stuff happens. While you're in bed, your mind puts on the weirdest, most amazing and sometimes scariest shows.

You might be swimming the Amazon River, hanging on to the wing of a plane, sitting down for a five-hour exam with your worst teacher or eating a pile of worms. Things that you know from real life, and perhaps haven't even paid much attention to, have a habit of cropping up in dreams in full Technicolor: the man who runs the newsagent suddenly has a starring role in a holiday you're dreaming of having taken in Zanzibar. A boy at school you never speak to turns out to be your best friend in a dream.

In the olden days, people believed that our dreams were full of clues about the future. Nowadays, we tend to think that dreams are a way for the mind to rearrange and tidy itself up after the activities of the day.

Why are dreams sometimes scary? During the day, things may happen that frighten us, but we are so busy we don't have time to think properly about them. At night, while we are sleeping safely, we can give those fears a run around. Or maybe something you did during the day was lovely but you were in a hurry and didn't give it time. It may pop up in a dream. In dreams, you go back over things you missed, repair what got damaged, make up stories about what you'd love, and explore the fears you normally put to the back of your mind.

Dreams are both more exciting and more frightening than daily life. They're a sign that our brains are marvellous machines – and that they have powers we don't often give them credit for, when we're just using them to do our homework or play a computer game. Dreams show us that we're not quite the bosses of our own selves.

HOW LONG WOULD IT TAKE TO WALK AROUND THE WORLD?

Rosie Swale-Pope

who ran around the world

I don't know how long it would take to walk around the world but it took me 1,789 days to run. I wore out fifty-three pairs of shoes!

I started the run for charity after my husband died, and I am so glad I did. It was amazing. I found out so much about people, and animals, and forests – and about myself.

One of my most unforgettable experiences was meeting a wolf pack in the forest in Siberia. Siberia is the loneliest place on earth. It's a winter fairyland of beauty and extreme cold.

I was in my tent at night when SUDDENLY I heard a noise. Moments later, a wolf put his head right inside the tent. Great furry paws stretching out in front of his nose, snow melting on his fur so it looked like he was wearing diamonds. Then he just backed away and was gone.

The wolf pack followed me at a distance for ten days, but never came close and didn't harm me. I remembered that wolves often look after people.

13

The people I met were pretty special, too. EVERYWHERE. Like the terrifying man in Russia who ran up to me waving an axe, and kindly gave me a parcel of bread! He was a woodsman called Alexei and he thought I must be hungry. Or like the children of White Mountain in Alaska who gave me a beautiful banner they had made before I set off into another thousand miles of wilderness. Their teacher said, 'We've named a star after you. When we look up at the night sky, we'll be thinking of you!'

At last I made it. All the way round the world. There are two footprints carved into the flagstones of my home in Tenby, Wales. My first step and my last step. There were twenty thousand miles in between.

Thank you for your great question. If you have a dream, whatever it is, GO FOR IT!! I wish you all the good luck in the world!

WHY DO WE HAVE MUSIC?

Jarvis Cocker

musician and broadcaster

That's a very good question. I wish I knew the answer (just joking!) Yes, it's true that if music disappeared from the world no one would die. It's not like air or water – we can live without it – but just think how boring life would be if it did disappear. Discos would go out of business, concerts would just be one big crowd of people staring at another, much smaller, crowd of people standing on a stage. In silence. As for Musical Statues . . . well, it would never get going would it?

But seriously, every society on Earth has music so there must be some point to it. In fact, some scientists think that humans were singing & making music long before they learned how to speak.

Perhaps it was our very first form of communication. & it can still be a way of communicating without words today: think about 'happy' songs & 'sad' songs. They both use the same musical notes (there are only twelve, you know) & yet are so different in mood. 'Ah, that's because of the words,' you might

15

say. But no. Try listening to the radio in a country where you don't speak the language. You'll still be able to tell the happy songs from the sad ones. It's the SOUND of the music that tells you. How does it work? I don't know – but it does. It's kind of magic & I think that's why we have it.

It's magic & we can have it whenever we want it. When you put one of your favourite songs on & get a sort of shivery feeling behind your ears & down the back of your neck (even goosebumps sometimes) that's one of the best feelings there is.

I like films & books & plays & paintings but they don't give me that same magical feeling. Only music does that.

& that's why we have it.

DO ALIENS EXIST?

Dr Seth Shostak

astronomer

When I was a kid, I would sometimes look up at the night sky with its thousands of stars and wonder: 'Could there be someone out there?'

Today, aliens – smart creatures that come from planets we've never heard of – can be found in a lot of movies and television shows. Aliens are everywhere, it seems. But not everything you see in movies or on TV is true. So what do scientists say about aliens? Do they exist?

The answer is: We still don't know.

Most scientists think it's possible that real aliens are out there. That's because the universe is so big. We live in a galaxy called the Milky Way. It's a very large group of stars, and we believe that our galaxy has about a thousand billion planets. In addition, there are at least a hundred billion *other* galaxies we can see with our telescopes. So the number of planets in the visible universe is about the same as the number of sand grains on all the beaches of Earth.

With so many places where aliens *could* exist, it certainly seems reasonable to believe that they really *do* exist.

How could we find them? Some people think that big-eyed visitors from another world have rocketed across space, and are flying around our skies in saucers. That would be very interesting, but most scientists don't think it's true. Why? Because the saucer reports are not convincing. When you see a light in the sky, there are many things it could be. For example, you might be seeing an aeroplane, a balloon or an orbiting satellite. Before scientists will believe that any of these mysterious lights are spaceships from another planet, they want better proof.

Another way to find aliens is to use big antennas to try and pick up radio signals coming from a faraway world. If we could hear a broadcast from another planet, we would know that someone's out there. Looking for these signals is my job, and so far, I haven't heard any alien shout-outs. But we've only begun to search. I think that by the year 2050, it's possible we'll find a signal.

Then we'll know the answer to the question 'Do aliens exist?' And the answer will be 'Yes.'

WHERE DOES WIND COME FROM?

Antony Woodward and Rob Penn

authors

Wind is just air moving from one place to another.

The source of the wind, as of so many things, is the Sun. As the Sun warms the Earth each day, it doesn't heat everywhere equally because some places catch the sunlight better than others. The place that catches it best is the Earth's waist, the equator, which is why places near the equator are hottest: the jungles and deserts and tropical islands. The places that catch the sunlight least well are the edges, the poles. This is why they are all snow and ice, and unless you're a penguin or a polar bear there's not much point being there.

Now, when air warms up, it *rises*. And as it rises – this is the important bit – something has to take its place: more air that's not so warm. As the warmed air rises, cooler air moves in to take its place, and – Presto! – that moving air is WIND.

Hurricanes and gales happen when the air is moving fast (because more air has risen, making room for lots more to rush in). Gentle breezes are when it's moving slowly, because less air has risen.

The atmosphere – the bubble of air around the Earth that we breathe – is warming and cooling and moving and mixing all the time, which is why our weather changes.

If it's all down to the Sun, how come the wind also blows at night? Because, of course, though it's night for *you*, it isn't night everywhere. Somewhere on Earth the Sun is always shining, warming, making the air move.

So there you are. As for the wind your dad produces? You know as well as we do: that's because he eats too many baked beans.

WHY DO WE SPEAK ENGLISH?

Professor David Crystal

language expert

If you travel some distance from where you live, you'll notice that people don't speak in the same way as you and your friends do. You'll hear different sounds – what we call differences of accent. And you'll also hear different words and ways of making sentences – what we call differences of dialect.

Accents and dialects show where you come from. People say such things as 'He sounds Welsh,' or 'She sounds as if she comes from London.' People from other countries have accents and dialects too. You can tell that someone is from America or Australia by the way they speak.

Accents and dialects change when people move from one place to another. They leave behind the way they used to speak, and start speaking in a new way. This is what happened thousands of years ago, as people began to explore our planet. When they settled in a new place, they would gradually develop new ways of speaking, and over a long period of time their speech would sound so different that if they went back to where they came from, they wouldn't be understood there any more. When that happens, we say that they had begun to speak a different language.

About three thousand years ago, groups of people who lived in southern and eastern Europe began to move into the northern regions that today we call Germany, Holland, Denmark, Sweden and Norway. They are known as the Germanic peoples, and the dialects and languages they spoke are all called Germanic too.

A monk called Bede wrote a book telling us how groups of Germanic people arrived in Britain in the fifth century from different parts of northern Europe. He says some of them were called Angles, some were called Saxons and some were called Jutes. They settled in different parts of Britain. And it didn't take long before they developed new ways of speaking.

After a while, people began to give these new settlers a name. They called them Anglo-Saxons – in other words, the 'English' Saxons, not the other Saxons who still lived in mainland Europe. They called this country 'English land', and eventually this became the name we now know, England. And the language spoken by these new Saxons, they called 'English'.

If you look at the English that the Anglo-Saxons spoke, you'll find it's very different from what we use today. There have been so many changes that it's almost like a foreign language. We call it Old English.

If you travelled back in time a thousand years, you'd have some difficulty understanding what the Anglo-Saxons were saying. But you'd recognise quite a few words that are still used in Modern English, such as *house*, *bed*, *child* and *friend*. And if you said to an Anglo-Saxon warrior, 'I live in that street,' he'd know what you were talking about, for all the words in that sentence have been in English for over a thousand years.

WHY DID DINOSAURS GO EXTINCT AND NOT OTHER ANIMALS?

Dr Richard Fortey

palaeontologist

Dinosaurs may have been big, but that didn't mean they could survive anything. Sometimes, being big is not such a good idea. Because they were large creatures, dinosaurs needed to eat a lot of food just to stay alive. The fiercest dinosaurs, like Tyrannosaurus, needed food that was other dinosaurs! If their lunch died out, then they would, too.

So when a great meteor – a huge rock – struck the earth sixty-five million years ago, so much dust and poison was thrown up into the sky that the Sun was blotted out. All plants need sunlight to grow. When the light stopped, the plants withered and died, leaving only their nuts and seeds to survive in the soil.

With no plants to eat, the great land-living vegetarian dinosaurs starved. After briefly gorging on the bodies of the vegetarian dinosaurs, the big hunters ran out of food, too, and soon died out like their peace-loving relatives. Now they are known only from fossil bones.

But other animals survived. Small mammals and snakes could live by eating beetles or other creatures that were protected in the soil. They were able to get through the disaster, though times must have been hard. Meanwhile in the sea, giant sea lizards died off, but crabs capable of eating almost anything were able to thrive. See how a crab will make off with a bit of your unwanted gristle next time you have a picnic at the seaside? They're not fussy. Lots of types of clams and snails with simple needs also survived.

Not everything that died out was enormous. A great group of coiled fossils called ammonites died out at the same time as the dinosaurs. Ammonites were underwater animals living in shells that looked a bit like curled-up sheep's horns, and they had been around for many millions of years longer than the dinosaurs.

Now here comes the surprise . . . Dinosaurs didn't really die out! Not all dinosaurs were huge: some of them were no bigger than a cat. Some of these small dinosaurs had feathers, and one of those feathered dinosaurs was the ancestor of the birds we see today. Birds can live on small treats and, if times are tough, they can fly to find a better place to live. Most scientists now agree that birds are descended from dinosaurs whose arms were modified into wings. And once you know this, you have to agree that they didn't really die out after all. The small ones simply flew away!

WHY DO CAKES TASTE SO NICE?

Lorraine Pascale

cookbook author and broadcaster

Do you know, I used to ask myself this question, many, many times over. The whole thing is like a big science experiment. You put eggs, butter, sugar and flour into a bowl, mix it all up carefully, put it in the oven and that is where the magic happens.

The ingredients form a magic web with each other. It is as if they all link hands together and then grow and grow in the heat of the oven. And while they are growing it is so hard to be patient because it also smells so nice.

I think that's the beauty of cakes and why they taste so nice. It takes a little bit of knowledge to put the cake together, but then the rest is magic. I mean, there are other things that use the very same ingredients, like pastry, but they never taste as good as cake!

Butter is a wonderful ingredient when used the right way, and sugar and eggs too. Then along comes the flour and just holds them all together and keeps them strong. It is all about having the perfect amount of these ingredients – to make the cake taste so very, very good that I'll have a big smile on my face when I eat it.

The beauty of this magic is that anyone can do it. My magic recipe starts with my favourite thing in the kitchen, my oven. And for the oven to perform the magic on the cake, it likes to be put on at 180 degrees centigrade. Then it's 200 grams of caster sugar and 200 grams of butter, creamed together with a big wooden spoon.

Then I add four medium-sized eggs and use that big wooden spoon to stir them all in, but for the cakes to taste extra good, I have to mix really, really hard.

Then the easiest bit: adding 200 grams of self-raising flour. I don't have to work so hard with this and can stir it gently. Now the cake needs to have something to sit in while it cooks in the oven, and its home is two round sandwich tins, twenty centimetres wide, lined with some baking parchment.

It's fun to watch the mixture as half goes into each tin and then into the oven. Then the part begins that I mentioned before: the magic spell in the oven.

The only thing with this is, if I try to take a peek in the oven and open it up before the magic time of thirty minutes is over, the cake refuses to go big and spongy and yummy. So I play the thirty-minute waiting game, which consists of lots of dancing and singing around the kitchen and then hey presto, the cake is ready and spongy and light!

Maybe this is why cakes taste so good – especially filled with jam and cream – because they're magic and making them is so much fun!

HOW DO PLANTS AND TREES GROW FROM A SMALL SEED?

Alys Fowler

gardening writer and broadcaster

I love seeds. I love that from an acorn an oak grows or that from the tiny pinprick of a poppy seed comes that wonderful, huge, blowsy bloom.

Not all seeds are tiny, though. Some are huge. The Coco de Mer is the largest seed in the world. It measures fifty centimetres and weighs up to thirty kilos, and although many have tried to give it appealing names, like love nut or sea coconut, it should really be called a baboon's bottom seed because that is just what it looks like! Other seeds are so small you can barely see them, such as Busy Lizzy seeds. Sowing those is hard work. One breeze and they all float away.

All seeds share the same basic principles whatever their size. Inside is a tiny baby plant, wrapped up in a seed coat to protect it. A seed is a little like a puzzle that you need a set of keys to open. The keys are water, heat and light (the last two both come from the sun). Once you have all the keys together, you can unlock the seed coat and the baby plant inside starts to grow.

29

The reason the seed coat is under lock and key is so the plant only germinates at the right time of year. Nobody likes getting out of bed in the winter and neither do most seeds. They sit in the soil waiting for the right temperature to kick life into action.

They need water to soften the hard seed coat so that the baby plant inside can break through. Think how tough a dry bean seed is, and then imagine a delicate little seedling pushing through. It can only do that if the seed has taken up enough water to soften the coat. Much like the way a flannel is hard when dry and you wouldn't want to wash your face with it. But once you've soaked it in water, it's nice and soft.

Each seed contains enough food for the little seedling to get going, so at first it doesn't necessarily need sunlight. (That's how it can grow underground.) But once it hits the surface of the soil, the sunlight gives the plant energy. So that eventually, with the right amount of water, heat and food, it can grow from a tiny seedling into a fully formed plant.

WHY DO MONKEYS LIKE BANANAS?

Daniel Simmonds

zookeeper at ZSL London Zoo

Monkeys eat lots of different foods. They eat fruit, vegetables, seeds, leaves and even insects. But they love bananas because they're very sweet and tasty. Just like us humans, monkeys enjoy eating yummy foods, and bananas are one of their favourite sweet treats.

Also, monkeys usually want to eat as quickly as possible, so that other monkeys don't take their food. (They can be quite naughty and will often steal each other's food.) Bananas are soft and squashy, so monkeys can eat them fast.

Different monkeys have different ways of eating bananas. The very greedy ones eat the bananas whole, skin and all. Others will peel off the skin and only eat the soft fruit inside. Some monkeys are not very good at peeling the strong skin, and instead roll the banana really hard until the soft part squeezes out of the ends. This is a clever but very messy way of eating bananas!

Monkeys use up lots of energy climbing, running and

31

swinging in trees. Bananas contain something called fructose – it's like sugar and gives the monkeys the energy they need to do all these things.

IS THE HUMAN BRAIN THE MOST POWERFUL THING ON EARTH?

Derren Brown

illusionist

Yes! All the amazing and powerful or terrible things we make or cause rely on our brains thinking them up in the first place. Brains allow us to have thoughts and language, and those then lead to great inventions, wars, medicine . . . anything you can think of that we come up with.

Brains let us sense the world around us. When we scrape our knees or see a flower, our knees or eyes don't really feel or see what's happening at all. The message has to go into our head to be worked out, and then it's our brain that makes us feel that the pain is in the knee or makes us see a flower in front of us!

Our brains also allow us to do something special that animals can't do, and that is *think about ourselves*. The fact that we can think about our own brains with our own brains is kind of weird but very clever.

What's really exciting is how our brains can play tricks on us. Like when you see a magician perform and you think something impossible has happened – the brain can trick us in ordinary life, too. We can get scared watching a film even though we're in no danger. Or we can think we've seen a ghost when we haven't. Or sometimes a mean person makes us feel bad. When that happens, we can start thinking 'I'm stupid, no one likes me,' or 'I'm worse than everyone else,' when that's not true at all . . . it's just our brain playing tricks on us!

When that happens we can imagine knocking on the tops of our heads and telling our brains to CALM DOWN. You see, our brains try to help and protect us but sometimes they really overreact, especially to bad stuff. Two really good ideas if that keeps happening are to talk about the problem (that calms our brains down nicely) and to find a hobby that we're good at (drawing, music, maths, magic, sport, anything) that we and our brains can enjoy together.

WHAT IS GLOBAL WARMING?

Dr Maggie Aderin-Pocock

space scientist

Today we hear a lot about global warming or climate change. As a space scientist I build machines that are helping us understand the changes that are happening. Yet if we look back into the past we can see that the climate of our planet has always been changing from ice ages to drought and heat waves, so why are we so worried now?

The problem with the climate change that we are experiencing today is that it is happening very fast. Faster than we have ever seen before. Also, the climate is changing not because of natural events like volcanoes and the sun's activity. Our climate is changing quickly because of things that we humans are doing. As we get more technologically advanced we need more power to run our many machines, like cars, aeroplanes and computers. My two-year-old daughter uses my iPad to watch videos, so we are starting young.

To get more power, we burn more fossil fuels like petrol for our cars or coal and gas to get more electricity. This gives us the power we want but it also produces 'greenhouse gases', such as carbon dioxide. These gases sit in our planet's atmosphere and trap heat from the Sun, changing our weather and generally increasing the Earth's temperature. This may not sound so bad but the rising temperatures lead to floods, drought and major devastation to people's lives across the world.

Is there anything we can do as individuals? This is a big problem that is affecting everyone on the planet but there are some things we can do that will make a difference.

Save energy: Climate change is happening due to our demand for more energy so anything we do to reduce this will help, like turning off lights when we are not using them, keeping our heating low and using low-energy bulbs.

Recycle whenever possible: It takes lots of energy to make materials such as cardboard, glass and plastic. By recycling we can save on some of that energy by reusing existing materials.

Eat locally produced food: If food is flown in from abroad, energy has been spent getting it to us. Eating locally grown food keeps transport energy low. I find this one quite hard. I love bananas and they are not grown in the UK so I have been trying to cut back on the number that I eat.

Tell others: This is a worldwide problem so the more people that are involved the better. We can all make a difference.

WHY DO I GET HICCUPS?

Harry Hill

comedian and former doctor

Hiccups are basically a twitching of the muscle that lies below your chest and above your tummy. This thin, trampoline-like muscle is at the base of the lungs, so when it twitches it causes you to take a little breath, which makes that hiccupy noise. The muscle is called your diaphragm (say 'die-uh-fram').

Most of us get hiccups after we've eaten too fast or drunk something cold or fizzy. The shock it gives our tummies makes the diaphragm muscle jerk sharply (like you would if you'd been given a shock). This pulls air into our lungs at such a speed that it rushes past the vocal cords in our throats, making the 'hic!' sound.

The diaphragm then twitches for a while, at random. The good news is that usually the diaphragm calms down in minutes. Though in one famous case, an unfortunate man in America had the same bout of hiccups for sixty-eight years!

Most doctors will tell you that this is the reason for hiccups. But I prefer another theory:

When we eat, the food is broken down in the stomach where it dies, releasing its ghost. Trapped within the stomach, this food-ghost is heard wailing and complaining about its fate. You can sometimes hear these noises as 'tummy rumblings'.

To live, the ghost must eat and indeed it does! (In fact, stomach ghosts are notoriously greedy.) Every time you have a shepherd's pie or some chips, your stomach ghost has a feast. As it eats, that food dies again releasing more ghosts who in turn eat food, die and so on, resulting in a stomach bulging with angry ghosts who just want a change of scenery.

Pretty soon this loose assortment of ghouls realise they can achieve a lot more by joining forces. They elect a leader and form a ghost collective thus becoming a sort of gastric SUPERGHOST! When this superghost grows big enough, it releases a series of baby ghosts, which appear as HICCUPS! Finally the superghost explodes in what we call a 'burp' and the whole process starts again.

That's what I heard, anyway. Which story do you think is true?

WHY IS SPACE SO SPARKLY?

Martin Rees

Astronomer Royal

Ever since we all lived in caves, human beings have looked up on dark nights and wondered about the twinkling points of light that we call the stars.

Our ancestors thought that the sky was like a huge dome over our heads – and the stars were attached to it, rather like lights on a huge Christmas tree. We have now learnt how absolutely immense the universe is, far bigger than our ancestors imagined. The stars are other 'Suns', each as big and bright as our Sun, but seeming so small and faint because they are much, much further away. The nearest star is so far away that even the fastest space rocket would take hundreds of thousands of years to reach it.

Astronomers have known for centuries that the Earth and the other major planets – Mercury, Venus, Mars, Jupiter, Saturn, Uranus and Neptune – are all in orbit around the Sun. You might wonder, therefore, if the other stars have planets orbiting around them, just as our Sun does. Until the 1990s, nobody knew the answer to this question. But astronomers have now discovered that most of the stars we see in the night sky do have planets around them. Some of these planets are the size of Jupiter, the 'giant' of our solar system. Others are the size of the Earth.

It's hard to see these planets, especially the ones no bigger than the Earth. They are millions of times fainter than the star they are orbiting around. The task is like looking for a firefly next to a powerful searchlight. But eventually astronomers will have telescopes that can get sharp enough pictures to see them.

You have learnt about the planets in our solar system. You have probably seen Venus and Jupiter, if not the others. But your children will find the night sky far more interesting. There will be a lot for them to learn about each star: how many planets orbit around them, how big they are, the length of their year, and so forth.

And this leads to the most fascinating question of all: is there life on any of these planets? If there is, will it be just bugs or insects, or could there, somewhere far away, be intelligent aliens? Could one of the planets be like our Earth, inhabited by people like us, who would regard one of these stars as their 'Sun'? Or are they very different from us? Perhaps creatures with seven tentacles, perhaps even computers and robots that have taken over from the beings that constructed them? Maybe some readers of this book will help us Earthlings to discover whether we are alone in the universe, or whether there is life among the stars.

One thing is certain. You will learn far more about the universe, and our place in it, than any astronomer knows today.

WHY CAN'T ANIMALS TALK LIKE US?

Noam Chomsky

linguist and philosopher

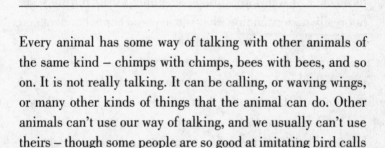

Every animal has some way of talking with other animals of the same kind – chimps with chimps, bees with bees, and so on. It is not really talking. It can be calling, or waving wings, or many other kinds of things that the animal can do. Other animals can't use our way of talking, and we usually can't use theirs – though some people are so good at imitating bird calls that the bird is fooled and thinks it's another bird.

Bees can tell other bees how far away a flower is, in which direction, and what kind of flower it is. They do it by a complicated dance that we would not be able to imitate. And it would be hard for us to give that kind of information as exactly as bees do. Monkeys have special cries that they use if they think a dangerous animal is coming towards them, or if they are hungry, or other things that they want to say. And other animals have something similar.

Human language is very different in many ways, and there is nothing like it in the animal world. Other animals just have a kind of list of things they can tell others about. They can't make up new ones. But humans can keep saying new things, things they haven't heard, and that maybe no one has ever said as long as humans existed. And you do it all the time, without even thinking about it.

Humans and other animals can talk to each other a little. If you have a dog, you can train it to sit when you say 'Sit!' and some other things, if you keep trying. And a cat can learn to meow if it wants you to do something for it. But they are not really understanding what you say, and cannot understand something new, the way another child can.

There are some birds that are very good at mimicking the songs of other birds, or even human words. Parrots can be trained to do that very well. It sounds a little like language, but they are not really using the language at all the way humans do. And like other animals, they cannot make up something new.

Some scientists who work with apes believe that they can be taught a little bit of human language. Others – like me – think that the scientists are fooling themselves and that the apes are doing something very different. It is an interesting question, and you might want to read and learn more about it. And maybe when you get older you may discover something new about it. There is a lot that we don't understand about human language and about animals.

HOW DO STORY WRITERS GET IDEAS FOR CHARACTERS?

Dame Jacqueline Wilson

author

I wonder how many characters I've made up writing my hundred books? It must be thousands and thousands. Imagine if they all came to life and crowded into my house for a party! I bet Tracy Beaker would be there at the very front, elbowing everyone else out the way so she could be first through the door. Hetty Feather would arrive in her plain brown uniform, and I'd take great delight in finding her a pretty party dress. Shy girls like Dolphin and Garnet and Beauty would hang back bashfully. Biscuits and Charlie might arrive with wonderful home-made cakes. Elsa would tell us endless jokes and Destiny would sing for us.

I can see them all – but of course they're not real. I made them all up. I rarely base my characters on real people. I certainly don't base them on myself. They spring out of my imagination.

Did you ever have an imaginary friend when you were very little? Did you pretend that your dolls and teddies were real,

43

and give them tea parties and put them to bed? Making up characters for stories is exactly like that. I might decide that I want to write a story about a girl stuck in a children's home, desperate to be fostered. Almost immediately fierce, funny Tracy pops into my head saying, 'I'm your girl! Write about me.'

You could make up your own characters too. Let's think about a girl who runs away from home. Why is she running away? Is she very unhappy – or simply naughty and wants an adventure? Is she clever and resourceful or will she panic? Is she big or little – plain or pretty – noisy or quiet as a mouse? What's she going to be called? Why don't you write a story about her?

HOW DO CARS WORK?

David Rooney

transport curator at the Science Museum, London

Cars move because their wheels are being spun round by an engine. When the wheels spin, their rubber tyres grip the road surface and the car moves. But how do the wheels spin?

Well, first we have to go to the filling station and put some fuel in the car. This is probably petrol or diesel, and it's like food for the car. It comes out of a hose we put into a hole in the side of the car. There's a tank for the fuel on the other side of the hole. You've probably seen this happen. It doesn't smell very nice.

Once the fuel is in the car and we start it up, the fuel gets sucked into the car's engine. This is the noisy complicated thing inside the front of the car. It burns a little bit of the fuel at a time to make a tiny little explosion, which makes a shaft rotate inside the engine. (A shaft is shaped like a pencil but made out of metal and much bigger and stronger – and you can't write homework with it.)

The trick with cars is to connect the very fast-rotating

45

shaft in the engine to the wheels underneath the car, so the car moves. This gets complicated, because the engine mostly likes to run very fast, but we might want the car to move either very fast or very slowly. So there's another machine in between the engine and the wheels, called a gearbox. It helps sort this problem out.

OK, so now we've got our car moving, but that's only the start of it. We've got to be able to turn it left or right depending on where we want to go. This is done by moving the steering wheel. This makes the front wheels point left or right, and the car goes the same way.

Now, it's all very well getting moving, but we have to be able to slow down or stop as well. This is done by the brakes. If you ride a bicycle you'll know that you slow the wheels down by pulling levers, which either push rubber blocks onto the wheels, or grip a metal disc attached to the wheel. It's pretty much the same with cars.

But next time you're being driven in a car, have a look at all the switches and levers and knobs and buttons the driver has to use. They're not all for making it go, steer and stop. There are loads of other bits that make a car work, from heaters to air conditioning, lights to locks, music systems to window washers.

When you think about it, cars are so complicated it's amazing they ever work at all.

WHY CAN'T I TICKLE MYSELF?

David Eagleman

neuroscientist

It's puzzling, isn't it? No matter where you try to tickle yourself, even on the soles of your feet or under your arms, you just can't.

To understand why, you need to know more about how your brain works. One of its main tasks is to try to make good guesses about what's going to happen next. While you're busy getting on with your life, walking downstairs or eating your breakfast, parts of your brain are always trying to predict the future.

Remember when you first learned how to ride a bicycle? At first, it took a lot of concentration to keep the handlebars steady and push the pedals. But after a while, cycling became easy. Now you're not aware of the movements you make to keep the bike going. From experience, your brain knows exactly what to expect so your body rides the bike automatically. Your brain is predicting all the movements you need to make.

You only have to think consciously about cycling if something changes – like if there's a strong wind or you get a flat tyre. When something unexpected happens like this, your brain is forced to change its predictions about what will happen next. If it does its job well, you'll adjust to the strong wind, leaning your body so you don't fall.

Why is it so important for our brains to predict what will happen next? It helps us make fewer mistakes and can even save our lives.

For example, when a chief fireman sees a fire, he immediately makes decisions about how best to position his men. His past experiences help him foresee what might happen and choose the best plan for fighting the blaze. His brain can instantly predict how different plans would work out, and he can rule out any bad or dangerous plans without putting his men at risk in real life.

So how does all this answer your question about tickling?

Because your brain is always predicting your own actions, and how your body will feel as a result, you cannot tickle yourself. Other people can tickle you because they can surprise you. You can't predict what their tickling actions will be.

And this knowledge leads to an interesting truth: if you build a machine that allows you to move a feather, but the feather moves only after a delay of a second, then you *can* tickle yourself. The results of your own actions will now surprise you.

WHO HAD THE FIRST PET?

Celia Haddon

author and pet agony aunt

We don't know the name of the person who had the first pet. But we do know that the first pet was probably a dog. Dogs began living with human beings thousands of years ago – some people think as long as forty thousand years ago. They may have just been like stray dogs, hanging around, following the human tribes that were hunting and gathering food. But perhaps some of them were treated as pets, companions that helped with the hunting.

One of the first pet dogs we know about was a puppy that was buried in a human grave about ten thousand to twelve thousand years ago in the country that is now Israel. In the same grave was a woman with her hand resting on the puppy's body, almost as if she was stroking it. Perhaps she wanted it as company in heaven or the next world.

Ancient Egyptians kept dogs as pets, too. We have pictures of them on tombs, sometimes with their names. They were called names like Ebony, Blackie and Big. The Romans had dogs as well and they gave the little pet ones names like Pearl, Dolly, Midget and Holdfast.

Cats probably began living near humans in Neolithic times when humans started farming. The first cat that may have been a pet was buried in its own little grave about nine thousand years ago, on the island we now call Cyprus. About forty centimetres away from the cat grave was a human grave. So the cat may have belonged to that human.

Ancient Egyptians kept cats as pets and we know the name of one of the first cat lovers. He was called Baket III, and he lived about four thousand years ago. On his tomb was a carving of a cat facing a rat! It's a very small cat or otherwise it's a very large rat, as they look about the same size.

Ancient Greeks and Romans also made carvings, paintings and mosaics of cats. Alas, they don't have names on them, and neither do most of the cats in Ancient Egyptian carvings. So we don't know what their names were.

WHY ARE PLANETS ROUND?

Professor Chris Riley

science writer and broadcaster

We've known that the Earth is round since 1519 when Portuguese explorer Ferdinand Magellan managed to sail right around it without falling off. And since then, of course, we've viewed it from space; first with satellites and then with people.

In 1961, Yuri Gagarin became the first person to fly right around the Earth, in just 108 minutes. And over the next decade twenty-four astronauts flew to the Moon, gazing back at their round, blue home planet with their own eyes from a quarter of a million miles away. Earth, the Moon, and every planet that we've explored in our solar system with our robotic probes is also round – or spherical.

To understand why all planets are round like this we need to go back in time. Way back to a time before the Earth or the Sun had formed. We find ourselves drifting in space, high above a vast cloud of gas and dust. The cloud is really big. So big, in fact, that we can't see the edges. And it's made up mostly of hydrogen and helium gases and a few other elements and chemical compounds.

As we accelerate time forwards again, we see a shock wave surging through the cloud. The shock wave has come from a neighbouring star, which has recently exploded at the end of its life. As the wave passes through the cloud it compresses the dust and gases, stirring them up, leaving vast, swirling patches behind.

These new rotating patches of gas and dust are slightly denser than their surroundings and they start to pull more material towards them. This pulling force around them is called gravity. The bigger these spinning clumps grow, the stronger their gravitational pull. They quickly grow in size, and some collide and are drawn together into even bigger spinning clumps. Growing gravitational forces pulling equally in all directions towards their centres ensure that these young planets quickly become spherical.

Now, living on a planet you'll have noticed that they aren't perfect spheres. Earth's mountains and valleys make its surfaces rather lumpy and bumpy. But you'll also have noticed that there aren't any mountains that stick right up into space. The force of gravity, pulling equally towards its centre, ensures that any mountains that might grow too big will sink back down again into Earth's hot, squishy interior, keeping the planet a more even spherical shape.

Well, almost spherical. Modern measurements of the Earth's size have revealed that it isn't exactly spherical. A planet's rotation throws out its equator against the force of gravity, resulting in a slightly squashed sphere. In the Earth's case this effect makes its diameter about forty kilometres wider at the equator than the poles.

CAN A BEE STING A BEE?

Dr George McGavin

entomologist

Yes it can. There are about twenty thousand species of bee in the world, but let's look at honey bees and bumble bees. Although some species are stingless, female bees typically have a sting to defend their colony against enemies that might steal their honey or even eat the bees themselves. Male bees do not have a sting and do nothing in the colony except for a few of them that will mate with the queen bee.

Honey bees will attack worker bees from other colonies if they try to enter, but a queen honey bee will only sting and kill other rival queens. A newly emerged queen will search the colony for cells where other queens are developing and when she finds them she will sting and kill them.

Bumble bees will attack workers from other colonies, too. They can sting them to death but usually they bite them and drive them out. In some cases the intruder may be able to hide inside the nest and may be accepted as a new member of the colony.

53

Bumble bees also fight and sting each other within the nest. The reason is complicated but is basically a way of reducing the numbers of males the colony produces. Why do they need to reduce the number of males? Because worker bumble bees can lay unfertilised eggs, which develop into males, but what the colony really needs are worker females.

The worker bees of some honey bee species have a special killing technique for large predators such as giant hornets. They form a ball around them and as the hundreds of bees vibrate their wing muscles the temperature and carbon dioxide levels inside the ball of bees increase and kill the hornet.

WHY DO WE COOK FOOD?

Heston Blumenthal

chef

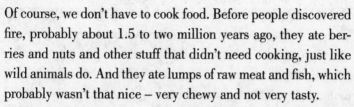

Of course, we don't have to cook food. Before people discovered fire, probably about 1.5 to two million years ago, they ate berries and nuts and other stuff that didn't need cooking, just like wild animals do. And they ate lumps of raw meat and fish, which probably wasn't that nice – very chewy and not very tasty.

The strange thing is, even after fire was discovered, we think that it was a long time – I'm talking many thousands of years here – before anyone realised that they could use it to cook! They built fires mainly to scare away wild animals. Some scientists think that one day, someone must have dropped a bit of raw meat or fish in the fire. After a while they noticed how good it smelled, tasted it and realised the heat had made it much more enjoyable to eat. Cooking was born, and eventually everybody was doing it because it has three very important effects on our food.

First, it makes a lot of raw, hard foods softer and easier to eat. Take a potato, for example. It starts off as a solid lump but cooking can turn it into soft, fluffy mashed potato. Second, it makes our food safer to eat. Sometimes food contains microbes that could make us ill. However, most of these microbes don't like really hot temperatures. Cooking them kills them, so they can't make us sick.

Third – and this is the most exciting effect for a chef like me – cooking can change food into something that looks, smells and tastes wonderful. Heat changes what it touches. Think of wood or coal in a fire turning to ash. Think of a candle gradually melting away. Heat not only makes the texture of food better, it breaks down ingredients into particles that are full of flavour, and it makes those ingredients react together to create new flavours. It can turn a pink, squidgy sausage into something browned, juicy and delicious. And it can turn a pale blob of dough into a lovely loaf of bread, which can even be cooked *again* to turn it into that crunchy, tasty piece of toast with your breakfast.

I've been cooking since I was a kid and to me it's still a kind of magic. It's amazing to watch it happen – and even more amazing to eat the results.

HOW DO YOU KEEP GOING WHEN YOU'RE LOSING IN SPORT?

Dame Kelly Holmes

athlete and winner of two Olympic gold medals

Firstly, it's important to know that everyone loses at something and actually it is OK to lose a match or race in sport. I didn't win all my races in my athletics career. In primary school I didn't always win but I loved taking part and always tried harder to be better in the next race.

The first big race I ran, when I was twelve, I came second. I was disappointed but it made me feel determined to do even better next time round because I wanted to win. It is also OK to feel disappointed as this just means you really want to do better.

Remember, if you don't win it is not always a failure. It is far more important to set goals. Before a race or training I would sit down with my coach and write down a goal: either a time I was going to try to run, or how I was going to race. It didn't matter how many people were in front of me. As long as I achieved the goal I set with my coach that was good enough.

If you focus on your goals you will get better every time.

It is also important to know that if you enter a race or a match knowing you or your team are going to win easily, you should challenge yourself and use it as a test so that you are better the next time you have more competition.

Winning doesn't happen overnight. You have to practise very hard and remember to also practise the parts you don't like to do. For example, when I was running I had to do lots of drills, which can get boring. But I knew they were helping me to run faster. The only time you lose is when you don't put one hundred per cent into your training and competition, because you can feel you have let yourself down.

The main thing, though, is: Don't forget to enjoy playing sport because that's why you play it in the first place!

WHY DO WARS HAPPEN?

Alex Crawford

war reporter

Wars happen because people don't talk enough to each other. I have interviewed militant fighters in Afghanistan who hate the West. The West is where I come from – and maybe you. It's the part of the world that includes Britain and America. Afghanistan is a country where British and American soldiers have been battling Taliban fighters for years. When the Taliban meet me they are astonished because I am often not only the first Westerner they have met but also the first Western woman.

When we start talking about our families and our children and what many people in the West think about them and the war, their whole attitude to me changes. We realise we are not so different and we probably want the same things. We both want peace.

Mostly wars happen because governments, which take decisions on our behalf, are frightened. A bit like when you are in the playground on your own because your mate is off school and the other 'gang' starts calling you names. What do you feel like doing? I bet sometimes you feel like calling them names back. And then when you get into a fight, it's hard, really hard to be the first one to stop and admit you're wrong. It's just the same with different countries.

WHY DO WE GO TO THE TOILET?

Adam Hart-Davis

author

Well I go when I want to go – and sometimes it gets absolutely desperate.

You need to wee, and you need to poo, for rather different reasons. You need to wee when your bladder gets full. Your bladder is like a floppy bag of skin inside the lower part of your tummy. Wee (or urine) collects in the bladder and fills it up, like a balloon fills with air when you blow into it.

When it gets nearly full it sends a warning signal to your brain, which makes you want to wee. The bladder is held shut at the bottom by an elastic ring called a sphincter (say 'sfincter'), which is like a tight rubber band round the neck of a balloon. When you go to the toilet you can relax the elastic and let the ring open. Then the urine can come pouring out.

In order to build up and repair muscles in your body, you need to eat some protein every day. It may come from eggs, or milk, or fish, or meat, or cheese, or beans. Your body breaks down the protein in these foods and builds up its own protein,

61

almost like a Lego construction kit. All these proteins contain an element called nitrogen, which you need for your muscles.

The problem is that you have to eat too much of it in order to make sure you have enough, and the extra nitrogen is a little bit poisonous, so you have to get rid of it. The way your body does this is to send it to your liver, where it is turned into a chemical called urea. As long as you drink plenty of water, the urea is washed down to your kidneys in your bloodstream. The kidneys filter out all the chemicals that can be recycled, and leave the urea dissolved in water, which is urine.

Birds can't drink lots of water, or they would be too heavy to fly. So they get rid of nitrogen by making uric acid instead of urea. Uric acid is a white solid, and that is why birds don't wee, but make poos that are partly white.

You need to poo for two main reasons. First you have to get rid of undigested fibre, which is made up of tough bits of plants. Everyone keeps telling us to eat lots of fibre, and packets of food have labels saying how much fibre they contain. You digest most of your food in your small intestine, which is a floppy tube about as wide as your thumb and five or six metres long. The intestine squeezes the food along, and the squeezing is made much easier if there are bits of solid fibre to push against. So although you can't digest the fibre, it helps you to digest the rest of your food.

You also need to poo to get rid of the remains of old blood cells. Blood cells take oxygen from your lungs all around your body, and this allows your brain and your muscles to work. The oxygen is carried by a chemical called haemoglobin. When the haemoglobin is past its use-by date, the blood takes it to the liver, which collects the bits that can be recycled, and sends the remains to be part of your poo. When haemoglobin wears out it turns into a brown chemical called bilirubin (say 'Billy Rubin'). This is what makes poo brown.

And that is why you go to the toilet.

WHY DO LIONS ROAR?

Kate Humble

wildlife TV presenter

You've got sport at school tomorrow and you can't find your trainers. You've looked everywhere: in the wardrobe, under your bed, on your windowsill because they were stinking out your bedroom. No. You pull everything out of your wardrobe, turn your school bag upside down, flip the dog's bed upside down in case the dog has stolen them. They've completely disappeared. So what do you do? You call your mum. 'Mum!'

She doesn't hear you. You try again, a bit louder: 'MUUUM!'

But she's in the kitchen washing up and singing along to the radio. So you take a deep breath, fill your lungs with air and shout at the top of your voice: 'MUUUUUUMMMMM!' and you watch her come running, slightly panicked because she thinks you've fallen down the stairs and broken both your legs. Which you haven't, obviously. But what you have done is get her attention. You've communicated with her.

All animals communicate with each other. Primates like monkeys and gorillas use similar methods of communication to us. As well as making a variety of sounds, they also use facial expressions and gestures. Ladybirds keep predators at bay by using colour. Their red-and-black wing cases act as a warning sign and say, 'Keep away, I'm dangerous.'

Dolphins click and squeak, but also splash the surface of the sea with their tails or leap clear of the water and bellyflop back down. A dolphin bellyflop is the equivalent of you going on Facebook and telling all your friends you've just heard the latest single by your favourite boy band and it's awesome. Except dolphins will be more interested in fish than boy bands. And lions? They grunt, moan, growl, snarl, hiss, spit, meow, hum, puff, and, of course, they ROOAAARR!

In the wild, most lions live in Africa, usually in the big, open grassy areas known as savannah. They live in groups called prides, usually one or two males and four or five females. Each pride has a territory, which the males in particular will defend to make sure no other lion comes in and steals either his antelope (which he likes to eat) or his females. Often these territories are very big and roaring is one way to defend them and let other lions know they are about to stray onto another male's patch.

If a male lion does come face to face with a rival lion he will roar to frighten him off. Roaring is also a useful way of keeping in touch with the rest of the pride – sort of like texting, only noisier. A lion that is really showing off could roar FIVE MILES away from your house and you'd still be able to hear him.

But a lion won't help you find your trainers. Only your mum can do that.

WHY DO WE HAVE MONEY?

Robert Peston

BBC business editor

Think of a world where there was no money. It would be very complicated. Let's say you wanted to buy a pizza. You would go to the pizza shop and order one. But remember that there is no money. So how are you going to persuade the pizza maker to give it to you? Well, the pizza maker, like you, needs and wants things. So perhaps the pizza maker would be prepared to swap a delicious pie for something you own or something you can make. But if you haven't got anything the pizza maker actually wants, then you wouldn't be able to have the pizza – which would be disppointing, wouldn't it?

Now think of it from the pizza maker's point of view. The pizza maker needs to get flour and tomatoes and cheese in order to make the pie. But in a world without money, how is the pizza maker going to persuade the farmer to supply her with the flour and tomatoes and cheese she needs? She could offer to swap pizzas for all those ingredients. But there are probably only so many pizzas the farmer wants, however delicious they may be.

That's why we invented money. Yes, we invented it. It didn't fall from the sky. It doesn't grow in the garden. We simply decided, thousands of years ago, that bits of metal would have a certain value, and that these bits of metal could be exchanged for all the things we want. Today money can be paper, or plastic, or electronic too. But the important thing about money is that it is something that we have all agreed has a value, so it can be swapped for the things we want.

The pizza maker is happy to take money from you in return for a pizza, because she knows the farmer will take that money for the ingredients, and the farmer knows she can use the money to buy what she wants (such as grain and fertiliser).

Money is one of the most amazing things we've invented, although no one knows the name of the brilliant inventor.

WHO WROTE THE FIRST BOOK EVER?

Professor Martyn Lyons

historian

It was so long ago that nobody knows. It's a mystery. But I can tell you about some of the very first books.

They weren't made of paper. In China, a long time ago, books were made out of wooden sticks from the bamboo tree. The sticks were tied together with string. Then people wrote on the sticks. The writing didn't go across from side to side, it went straight down, from the top to the bottom.

The first person to make paper was called Cai Lun. He was a Chinese man who wore a long robe and tied his hair in a pigtail at the back of his head. Cai Lun made paper from rags and old clothes. So if you threw away your T-shirt, Cai Lun could make it into a notebook. I was joking about the T-shirt, by the way. Don't throw it away, just put it in the washing machine.

Chinese people thought that an old man called Confucius was very wise. They wanted to write down everything he said so they would remember it. All Confucius's words were carved onto fifty enormous stones, each stone as big as a man. It was the heaviest book ever. It took eight years to write and two hundred people to carry it.

The very first big library ever was in Egypt. The books in the library didn't have pages – they were written on rolled-up paper, called scrolls. Imagine a whole library of books that looked like giant toilet-paper rolls. One day the library caught fire and all the books were burned. How terrible! Please don't ever let something like that happen to your favourite books.

WHY DO ELEPHANTS HAVE TRUNKS?

Michaela Strachan

wildlife TV presenter

Because they would look silly with a glove compartment! On a more serious note, elephants have trunks for many different reasons. It's really quite incredible what an elephant can do with its trunk. It can use it for eating, drinking, showering, hugging, touching, smelling, swimming, pulling down trees, picking things up and fighting.

In fact, no animal has a better and more useful nose than an elephant. Its trunk is its nose and upper lip all in one. It's strong, supple and delicate. Just imagine if our arms could do as many things as an elephant's trunk. I may be able to touch, pick things up and hug with my arm but I certainly can't smell, shower, or sniff up water.

An elephant's trunk is really strong. Strong enough to pull down a tree! It can also be incredibly delicate. Delicate enough to pick up a pencil or a peanut. It's really long, so it can reach up to the top of trees to get leaves, or reach down to suck up water to squirt into its mouth to drink, or to hose itself for a refreshing shower. It can also squirt dust over itself as protection from biting flies.

Have you ever seen an elephant swim? If the water gets too deep it uses its trunk as a snorkel. How cool is that? Wish I could do that with my arm! Elephants have long trunks because they're giant animals with long legs and a huge head. Their trunks are vital for feeding, with forty thousand muscles and tendons and an extremely sensitive tip.

It takes an elephant a good year to work out how to use its trunk properly and it can be quite amusing watching a baby elephant learning how to co-ordinate all those muscles. I've even seen elephants using their trunks to paint! Obviously these were in captivity, but the result was a work of art and a trunkload of fun. (Groan!)

WHY ARE SOME PEOPLE MEAN?

Dr Oliver James

psychologist

You know what it's like when your mummy or daddy gets annoyed with you, when it wasn't your fault? It makes you really, really angry, as well as a bit sad too?

Well, maybe afterwards you go and do something annoying to another child. Perhaps it's your brother or sister, and you know how much it winds them up if you hide their favourite toy or remind them how useless they are at maths. Or maybe it's someone at school, and you know how to make them angry, like by telling them it's fish for lunch when you know they hate fish, or by calling them names.

That's why people are mean. Somebody has done something to them that made them angry or sad. And they want to get rid of that feeling. So they try to make *you* angry or sad. It's like using other people as a dustbin for rubbish. They have this rubbish feeling, so they try to dump it in you. For a little bit afterwards, they feel relief. They think, 'Thank goodness I got rid of that rubbish.'

73

But it doesn't really work. After a while the rubbish has a funny way of reappearing, like when you throw something in the sea or a pond and it comes floating back up to the surface. They might feel bad about having been mean. They might have a nasty dream about it or feel sort of snarly and grrrrr. Or they might feel sad and cry.

Maybe they don't feel that bad about it. And because they are mean to so many people, no one likes them very much. Which makes them even angrier and sadder. So they dump even more rubbish in the people around them. And it just gets worse. They end up feeling as if they're in the middle of a rubbish dump.

The next time someone is mean to you, just ask yourself this: 'I wonder why that mean person is so unhappy? I wonder what is making them so sad or angry that they're being mean to me?'

The odd thing is, if you do that, you won't feel nearly so bad.

HOW DO TREES MAKE THE AIR THAT WE BREATHE?

Dr David Bellamy

botanist and conservationist

All the trees, plants and animals that share this wonderful world with us need three invisible gases to grow and keep healthy. These magic gases are called carbon dioxide, water vapour and oxygen. They are the main building blocks of all living things and without them there would be no life on Earth.

Every time you breathe in, you fill your lungs with fresh air that contains oxygen. Your body needs a lot of oxygen to keep going, so it is rapidly used up and is replaced by carbon dioxide. When you breathe out, this carbon dioxide is flushed into the open air.

All plants, including trees, gather carbon dioxide and water vapour from the air. They use energy from sunlight to change these gases into sugars and other basic foods to help them grow. As they do this, they release oxygen into the air.

This process is called photosynthesis and it is the only source of sugars and oxygen for all living things.

People and plants breathe in different ways, of course. We

have noses and mouths through which we take in oxygen. These are connected to our lungs, which pump the life-giving gases in and out. Plants don't have lungs but they have lots of breathing holes scattered over their leaves and stems, which let the gases in and out. These are connected by a hidden plumbing system of very thin tubes that carry water from the plant's deepest roots in the damp soil, up to the highest leaves.

All plants do their best to keep their pipes full of water. But when the leaves get too hot or the soil too dry, they close their breathing holes to save water. When the breathing holes are open, though, water evaporates from the holes. And at the same time carbon dioxide moves into the plant.

When I'm working in the garden I like to sing because I know that all the plants are saying thank you for the carbon dioxide that I'm breathing out! Of course, I can't hear them. But I know that the carbon dioxide is helping them grow more flowers, fruit, cereals and vegetables.

Making life from invisible gases and sunbeams may seem like a fairy story. But it's happening all around the world, and all around you and me. I'm very glad this happens because if it stopped, I wouldn't be here to answer your super question.

IF THE UNIVERSE STARTED FROM NOTHING, HOW DID IT BECOME SOMETHING?

Dr Simon Singh

science writer

Scientists have discovered evidence that suggests the universe was created after a giant explosion called the Big Bang. All the tiny bits that make up today's galaxies, stars and planets suddenly appeared out of this explosion. In fact, the Big Bang also created space itself. Even more bizarrely, the Big Bang also created time.

Because of the explosive nature of the Big Bang, the universe has been expanding ever since it was created. This means that the galaxies have been flying apart and will get further apart in the future. However, the force of gravity might change all of this.

Gravity is an attractive force, which means that it tries to pull everything together. That is why when you fall, you fall down to Earth and you do not fall up, away from the Earth. Gravity is pulling you and the Earth towards each other. Gravity means that every bit of the universe is attracted to every other bit.

So it is possible that, in the distant future, gravity might slow down the expansion of the universe, stop it and reverse it. That would mean that the universe would begin to contract.

Then, in the very, very far future the universe would have the opposite of a Big Bang, sometimes called a Gib Gnab (Big Bang spelt backwards) or a Big Crunch. This might lead to a Big Rebound and another Big Bang and so on. The history of the universe would be Big Bang, Big Expansion, Big Halt, Big Collapse, Big Crunch, Big Rebound, Big Bang . . .

In other words, the universe did not start from nothing, but rather it started from the collapse of an earlier universe. Our universe is a recycled version of a previous universe.

Unfortunately, we do not have much evidence to prove that this recycling universe theory is correct. Indeed, there is some evidence that indicates that the universe cannot reverse its expansion. Therefore, scientists continue to explore this mystery.

While we wait for a scientific answer to this question, it is worth mentioning St Augustine, a fourth-century Christian philosopher, who was faced with a similar puzzle. Instead of the question 'What came before the Big Bang?' someone asked him, 'What was God doing before he created the world?' He replied that God was creating hell for people who asked questions like that.

WHY DO PEOPLE HAVE DIFFERENT-COLOURED SKIN?

Carl Zimmer

science writer

Let's start by taking a look at how our skin gets its colour. In your skin, you have special cells that make dark clumps of molecules called pigments. Different clumps have different colours. Combinations of the clumps can produce other colours. And the more pigment made in the skin, the stronger its colour will be. Very pale people from Sweden might make very few pigments in their skin. Very dark-skinned people from Senegal in Africa make many pigments.

To know why people have different skin colours, we need to look at the good things that pigment does for us. Skin pigment can act like a natural sunscreen. Sunlight contains dangerous kinds of energy that can cause sunburn, and can even cause a disease called cancer. When dangerous sunlight hits the skin, the pigment can grab it and keep it from harming a person. In Africa, where the sun is very intense, dark skin is like a strong shield protecting people from cancer.

But if we didn't get any sunlight, we'd get sick in a different way. We need sunlight to help us make something called vitamin D, which our body needs to stay healthy. In Africa, there's so much sun that a little sunlight can get into dark skin. In a place like Europe, where the sun isn't so strong, a dark skin might not get enough sun to make vitamin D. That's why people whose ancestors are from Europe have lighter skin. Europeans with pale skin don't get more skin cancer, because there's less sunlight in Europe.

WILL THE NORTH AND SOUTH POLES EVER MELT COMPLETELY?

Dr Gabrielle Walker

writer and broadcaster on climate change

The North and South Poles are surrounded by ice, and that ice might melt in the future. To understand why, it's probably best to think of the north and south separately.

The North Pole is a point at the 'top' of the world and all around it is a very cold ocean. There are lots of fabulous animals at the North Pole, like polar bears and whales and big fat walruses with moustaches and very long tusks, and they all live in and around the water.

Because it's so cold, the top part of this polar ocean is frozen solid, especially in the winter. Although this ice is quite thick, enough that you can sometimes drive on parts of it with special ice-scooters or tractors, it can still melt fairly easily when summer comes. In fact that's already happening. Because of global warming, the sea ice in the north has been melting away for decades and in some summers there is only half an ice cap there instead of a whole one! That's why so many people are worried about the fate of the polar bears, and also whether we humans might start to suffer from the warming too.

The South Pole is a bit safer, because the ice there is much, much thicker. Instead of being a frozen ocean, it's a gigantic frozen land called Antarctica, and in the middle the ice is so thick that you're walking on an ice mountain more than two miles high.

Around the edges of Antarctica, you can find lots of penguins (and they really are as cute as they seem). But in the middle it is so cold and the ice is so thick that there is no life at all – unless you count the human scientists who go there to study the ice and snow.

There is even a real pole stuck into the ice at the South Pole by American researchers who have a permanent base there. It looks like a barber's stripy pole and you can have a picture taken next to it. Or better still, if you do a handstand on top of the pole, get someone to take a picture, and then turn the picture upside down, you can look like you're hanging off the bottom of the world!

But we now know that even the ice of Antarctica is melting – especially at the edges – and one day it could all be gone. That wouldn't be very good for us humans because the melting ice makes the sea rise up higher, which could be a problem for all the people living near the seaside all around the world. But it might be good for Antarctica, because creatures would then be able to live in the interior where it's now too cold. A hundred million years ago, the whole world was so warm that dinosaurs lived in steamy swamps at the South Pole! If the ice melts again who knows what will be living there next?

WHERE DOES 'GOOD' COME FROM?

A. C. Grayling

philosopher

We use the word 'good' to describe things we like, and things that make life better, and kind things that people do for other people. We describe people as good when they are honest and nice to others, when they keep their promises, and try their best. Goodness is very important because it really does help to make our world a better place.

Ever since people first asked themselves, 'What is the right way for us to behave and to treat one another?' there has been discussion about the nature of goodness. The ancient Greek philosophers began a debate about goodness that has lasted ever since. They taught us to see that goodness is not only about the things we do, but also the way we think. This means that our attitudes are important, because our actions come from our attitudes, so thinking about the right way to live and act is something we must all do.

So we must all ask ourselves: What do I think is good? Why

do I think that? I am just about to do something: is it right or not? When you answer these questions you must be sure that the answer will persuade other people too: it is too easy just to persuade yourself!

Thinking about goodness so that we can do good things involves talking to other people, learning about what different societies think and why they think it, and asking for the reasons that people have for thinking something is good or bad.

What we learn from all this is that 'good' comes from responsible and sensible thinking about the effect that our thoughts and acts have on us and on others, and on the world around us.

WHY IS THE SUN SO HOT?

Dr Lucie Green

space scientist

Why the Sun is hot has baffled people for thousands of years.

One early idea was that the Sun is a burning lump of coal, but today we know that the Sun is made up mostly of hydrogen particles and wouldn't burn in the way that coal does. Instead the hydrogen in the centre of the Sun is being squeezed together so much that it actually starts to stick together and form a gas called helium.

Work by Albert Einstein helped scientists understand that when these particles are squeezed together they can release enough energy to keep the Sun shining and keep it hot. The temperature in the centre of the Sun is fifteen million degrees centigrade but the temperature at the surface is much lower at 5,700 degrees centigrade. Water in a kettle boils at one hundred degrees centigrade, so just try to imagine how hot the centre of the Sun is.

Today we can study our Sun in great detail using telescopes in space and we see that the Sun has a surprisingly hot atmosphere, much hotter than the surface, with a temperature of a million degrees centigrade. This is surprising, as the heat coming from the Sun's surface cannot produce an atmosphere this hot. The hot gases in the atmosphere shine brightly in X-rays and ultraviolet light. Pictures of the Sun taken with telescopes in space that can see the X-rays and ultraviolet light have helped us understand that the atmosphere is kept hot by immense magnetic fields running through these gases.

Using spacecraft such as SOHO, the Solar Dynamics Observatory and Hinode, we know that these magnetic fields are constantly moving and rippling, and that the magnetic fields power explosions and heat the gases in the Sun's atmosphere to a million degrees.

WHAT IS THE MOST ENDANGERED ANIMAL IN THE WORLD?

Mark Carwardine

zoologist and conservationist

Until recently, the Pinta Island giant tortoise was the rarest animal we knew about. There was just one survivor and his name was Lonesome George. He lived in the Galapagos Islands, a long way off the coast of South America, and was believed to be about a hundred years old. But he sadly died in June 2012 and so his sub-species is now extinct.

We also know that the most famous of all endangered animals, the giant panda, is *not* the most endangered animal in the world. It is certainly rare – experts think there are only about 1,600 giant pandas living in the bamboo forests of China – and its population is probably declining. But many animals are even more endangered than that. Some have already disappeared from the wild, although they are not officially extinct because there are still some survivors in captivity. One of these is a kind of parrot, called Spix's macaw, which has a population of about 120 – all living in zoos or being kept as pets.

Other animals have larger populations in the wild but they face huge threats and are therefore even more endangered. These include lots of well-known species, such as the Javan rhino, the tiger and the mountain gorilla. But they also include the vaquita (a little porpoise living off the coast of Mexico), the greater bamboo lemur (a monkey-like animal from Madagascar), the addax (an African antelope) and many other animals that most people have never even heard of.

Altogether, there are more than two thousand *critically* endangered animals in the world – they're just the ones we know about – and many thousands of others that are simply endangered. But they are not necessarily all doomed to extinction. The grey whale is a great example: after commercial hunting was banned in 1946 its population bounced back from a couple of hundred to a much healthier twenty-one thousand. So the good news is that, if we try hard, conservation efforts really can save endangered animals from extinction.

WHY DO GIRLS HAVE BABIES AND BOYS DON'T?

Dr Sarah Jarvis

medical doctor and broadcaster

In some ways, girls and boys are quite similar on the outside. After all, they have the same number of arms, legs, ears and noses. One of the biggest differences between women and men (apart from the bald patch you sometimes see on men!) is that women have breasts and men don't. Also, men have penises while women don't. Inside our bodies, women and men also have some bits the same and some different. They both have hearts to pump blood round the body. They both have lungs to breathe with. But inside a woman's tummy area is something called a womb. It's usually about the size of a chicken's egg, but it can blow up like a balloon. It's hollow and it has a soft lining. Men don't have wombs.

A baby comes from a woman's egg and a man's seed. Together, they grow into a baby, and that's complicated. Babies need to get all their food from their mummies before they are born. Inside the womb, they can attach to their mummy and get everything they need to grow from her body.

Babies also need to be protected. When babies are first born, they can't do much except feed, cry and sleep. But by the time they're born they are nine months old. Before that they can't breathe on their own.

When a woman is pregnant, the baby in her womb floats in a fluid and doesn't need to breathe. But it does need to grow. Because the womb is so stretchy, the baby can grow from the size of a pea to the size of four bags of sugar before he or she needs to come out.

Of course, the differences don't stop when the baby is born. When a woman has had a baby, her breasts make milk, which has everything the baby needs to help it grow. Daddies are great at many things, but they can't have babies!

IN VICTORIAN TIMES WHY DID KIDS DO ALL THE WORK?

Claire Tomalin

author

Grown-ups did a lot of work in Victorian England – of course! – as engineers building railways, as scientists, as factory workers, as teachers and writers, doctors and nurses. But children worked every bit as hard. They did dangerous jobs: some were sent down mines, other very small boys were forced to sweep chimneys. Although there were several Acts of Parliament intended to stop this, they were ignored for years and children went on being forced up chimneys until 1875, when the death of a boy trapped halfway up one finally brought an end to such cruelty. Children also worked long hours in factories and mills.

Rich children went to school, poor ones rarely did. Many children were born in the workhouses set up for the poor. Charles Dickens, the great storyteller of Victorian times, describes in his novel *Oliver Twist* how badly they were treated there. To begin with they were half starved – everyone knows about Oliver asking for more, and being punished – and at only nine or ten they were sent out to work as servants or apprentices. Oliver begs not to be apprenticed to a chimney sweep. Later he is kidnapped by criminals who try to make him into a pickpocket, and force him to climb through a small window into a house they want to burgle.

Dickens knew a great deal about child workers. He was himself set to work at twelve years old in a factory putting boot blacking into jars. He hated doing it and longed to be at school. In his stories he describes the street children of London. The boy Jo who can't read or write keeps himself just alive with pennies he is given for sweeping a street crossing, and dies young. Then there is Charley, an orphan girl who manages to feed and lodge her little brother and sister by going out to do washing, leaving them locked up for safety while she works. Another girl, Jenny, who is disabled and can hardly walk, earns her living making dolls' clothes. Circus people trained their children to perform as acrobats and riders. Theatre people would put a daughter on stage as soon as she could walk and present her as the 'Infant Phenomenon'.

Some good people set up 'Ragged Schools' for the street children, where they were given a little teaching. They were often absent and on returning explained they had been in prison. Dickens describes a small boy pickpocket being very cheeky to the judge when he is in court. And although Dickens had little education himself, he grew up none the worse for having been a child worker, and became a world-famous writer. Today English children are given education, but in other parts of the world children are still forced to work long hours of hard labour.

WHAT IS GRAVITY, AND WHY ISN'T THERE ANY IN SPACE?

Dr Nicholas J. M. Patrick

NASA astronaut

Gravity is a pulling force that every object in the universe exerts on every other object. And there's plenty of it in space!

The bigger and closer an object is, the harder its gravity pulls. The Earth is very big and very close to you, so it exerts a lot of gravitational force on you, holding you down to the ground and preventing you from drifting off into space. We call this force your weight. Everything else exerts a little gravitational force on you as well: the Moon pulls on you, for example, although not enough for you to notice. The Moon also pulls on the Earth's oceans, causing the tides.

But gravity doesn't just exist here on Earth – it fills space. Within our solar system, our enormous Sun's gravity reaches out and holds the Earth and the other planets in their orbits around it, just as Earth's gravity holds the Moon in its orbit.

So, if Earth's gravity can reach out to the Moon and beyond, why don't astronauts *feel* it when they're orbiting the Earth in a spacecraft? Why do you feel 'weightless' in orbit?

The answer is – somewhat surprisingly – that when you are in orbit, you are actually falling towards the Earth because of the pull of Earth's gravity. And since you are falling, you're not standing on anything, so you don't feel your weight on your feet and legs. The reason you never hit the ground when you're in orbit in a spacecraft is that you are falling *around* the Earth. You're travelling forwards at 17,500 miles per hour, so quickly that the curved Earth falls out from under you as quickly as you fall towards it.

As an astronaut, I've experienced weightlessness for weeks on end, living aboard the space shuttles *Discovery* and *Endeavour* and the International Space Station. When we're not working, we enjoy the view, and practise our floating technique. With a little practice, you can float motionless in the middle of the station for several minutes, until the slight breeze from the air-conditioning fans blows you gently towards a vent!

WHY CAN'T WE LIVE FOREVER?

Richard Holloway

author and broadcaster

If we lived forever and no one ever died, within a few years the world would be so crowded that we wouldn't be able to move and play and run about.

It would be like having more and more people come to live with you, without making any more space for them in your home. Fun at first but soon you wouldn't be able to lie down or have your own bed or play with your own games, it would be so crowded!

And we would soon eat up all the food in the world because there wouldn't be enough for everyone, and we would get very hungry and very sick and probably start fighting over what there was to eat.

Worst of all, life would get terribly boring and tiring. It would be like going to a school that never had playtime or holidays. It would just go on and on and on, with the same things happening over and over, for ever and ever.

Because we don't live forever, we can look forward to growing

95

up and having children and then growing old and dying, and leaving space for our children to live and grow and have children, and on and on, forever.

HOW DOES WATER GET INTO THE CLOUDS SO IT CAN RAIN?

Gavin Pretor-Pinney

author and founder of the Cloud Appreciation Society

Clouds are made of millions and millions of tiny bits of water. Sometimes they are little droplets, sometimes tiny crystals of ice. It does seem strange that this water should just appear in the sky when we don't see it getting up there but remember: just because you can't see something doesn't mean it's not there.

Sometimes water is invisible. Not runny water, of course, like the stuff you drink. We can all see that. Nor water when it is frozen solid into ice, which is also easy to see. The form of water that you can't see is when it is a gas. This is when the tiniest bits of water, called water molecules, are flying around separately in the air, rather than being stuck together into runny water or hard ice.

When water is a gas, the molecules are zooming around with lots of space between them. And since all molecules are far, far too small for us to see, water is invisible when it is a gas like this. Only when thousands of millions of water molecules stick together to form a tiny droplet do we stand any chance of seeing them. And that is exactly what happens when a cloud forms up in the sky.

You may not realise it, but there is a lot of this invisible water around. It is part of the air we breathe. The water molecules get into the air by lifting off the top of oceans, snow, puddles and any other water down at ground level. Although they are too small to see zooming around on their own, the molecules are very much there, bumping into each other along with all the other molecules of the air.

The warmer the air is, the more water molecules lift up into it and the faster they fly around. But how does this invisible form of water get up high in the sky and turn into a white, puffy cloud?

The air in the lower few miles of our atmosphere is very swirly, and there are all manner of ways that the air near the ground can lift up into the sky. It might rise as the wind blows over a mountain. It might float upwards as it is heated by the sun-warmed ground. No matter how it gets up there, it always cools as it rises. And this is what makes clouds appear.

As the air cools, those invisible water molecules don't zoom so fast. If it cools enough, they start to stick together into droplets when they bump into each other. When air gets colder as it rises, lots of these droplets can appear and grow large enough for us to see as a white cloud.

If the air keeps on rising and keeps on cooling, the cloud's droplets turn into tiny pieces of ice. These can grow large enough to start falling back downwards again and land as snow or rain.

WHY DO ANIMALS THAT FLY HAVE FEATHERS, NOT INCLUDING BATS?

John 'Jack' Horner

palaeontologist

Actually, the only living animals with feathers are birds. And although they use some of their feathers for flight, most of their feathers are for other purposes. When we look at fossils we see that little dinosaurs were apparently the first animals to have feathers, and they did not use them for flight. The feathers of these little dinosaurs were primarily for insulation and showing off. Showing off is called 'display' and we see that male birds use feathers to show off to females and sometimes to other males. Showing off is how animals attract mates.

Over the past few years scientists have come to the conclusion that dinosaurs gave rise to birds, which means that dinosaurs are the ancestors of birds. Dinosaurs actually invented most of the features we think of as characteristics of birds, like feathers and hollow bones and the wishbone and hard-shelled eggs.

In other words, there are so many characteristics shared between dinosaurs and birds that we palaeontologists now classify birds within the group *Dinosauria*. Birds are living dinosaurs! And because birds are a type of dinosaur, I have started working with some biologists to try to recreate a dinosaur from a bird by turning on and off certain genes in their DNA. Using chickens, we are looking for the genes that will allow them to grow long tails and have long arms with hands instead of wings. We are also trying to grow a chicken with teeth.

If we are able to create a bird with dinosaur characteristics we will call it a Chickenosaurus or a Dinochicken! Once we can make a dinosaur out of a chicken we will be able to make a dinosaur out of any bird because all birds are related. Some kids would like us to make them out of ostriches so we have big ones, but I think we should keep them small so they don't eat us. What do you think?

HOW DOES MY BRAIN CONTROL ME?

Baroness Susan Greenfield

neuroscientist

There are two important words in this question, 'brain' and 'me'. We first need to make sure we really understand what they mean.

The brain is a sludgy thing that fills up the inside of your head and looks a little like a very large wrinkled walnut. Although unlike a nut, it's soft, like a soft-boiled egg. But it does much, much more than a nut or an egg: it enables you to see, to hear, to feel, to smell and taste. It's also the central headquarters of your body for directing all the many different muscles in your arms and legs so that you can move. Most importantly of all, your brain is what you think with, so eventually you can think about being 'you'.

Let's see what happens inside your head . . .

When you are a newborn baby, your brain is the same size as a baby chimpanzee's. But then something amazing happens. There are about a hundred billion tiny building blocks ('cells') that can only be seen under a microscope and it is these cells that make up your brain. However, after you are born these cells in the human brain start to make stringy connections with each other, and as the connections lengthen and increase, so your brain grows accordingly, way more than for a chimp.

Why is this interesting or important?

We humans don't run particularly fast, we don't see particularly well, and we're not that strong compared to many other animals. But we can live and thrive across more of the planet than any other species, because we do something far better than any other. We learn.

It is because we are so good at learning from experience that we can adapt to any environment in which we are born. And we are good at learning because our brain cells are fantastic at making connections every moment we are alive. Every experience you have will change your brain connections. So even if you are a clone – an identical twin with the same genes as your brother or sister – you will have a unique pattern of brain-cell connections because only you will have a certain set of experiences. Even if you live in the same house with the same family, individual and unique things will happen to you that are different from what happens to everyone else. Every time you do something ordinary like talking with someone, playing a game, eating a certain food or looking out of the window, your brain-cell connections will adapt in a unique way to make you the extraordinary individual you are.

The answer to the question, therefore, is that 'my brain' and 'me' are the same. So one cannot control the other.

However, how the feeling of being you can be caused by something that looks like a nut and feels like an egg is one of the hardest and biggest puzzles still to solve.

HOW DO CHEFS GET IDEAS FOR RECIPES?

Gordon Ramsay

chef

Chefs get ideas from all sorts of places: old cookery books, family, friends and other chefs. But I am mostly inspired by what I find in the markets.

I love visiting farmers' markets early in the morning when the stalls are just opening. The vegetables, fish and meat are all fresh and I get to see some fellow chefs negotiating prices of ingredients for their restaurants. The best bit is spending a few minutes discussing the food and where it came from with the people who produced it. They are truly passionate about offering quality food that has been farmed in an organic and sustainable way – without harming the environment.

Once I have selected my seasonal ingredients, I start to develop recipe ideas and flavours that will bring out the essence and complement what I have just bought.

Each season offers a new range of produce, taking me back to some trusted old recipes and developing new ones. In the cold wintery months, there's nothing better than a hearty stew.

Carrots, parsnips, celeriac, squash, turnips and potatoes make a great base for this warm winter dish.

In spring, asparagus is in season and this is one vegetable that always puts a smile on my face and prompts a whole new burst of kitchen activity. I love roasted lobster tail with British asparagus, morel mushrooms, lemon and vanilla sauce!

In the summer, berries are at their best. One of my favourite things to cook is a lemon tart with fresh summer berries; the sharpness of the lemon is softened by the sweetness of the fruit. Delicious!

Pears hit their peak in the autumn months and I can't wait to cook pear tarte Tatin. The spices star anise, cardamom and cinnamon all bring a pear to life, and it's an easy dessert for a special occasion.

It's important to have fun when you're cooking. Experiment with different flavours to create new dishes. You just never know what delicious and interesting culinary adventure you could have.

ARE WE ALL RELATED?

Dr Richard Dawkins

evolutionary biologist

Yes, we are all related. You are a (probably distant) cousin of the Queen, and of the president of the United States, and of me. You and I are cousins of each other. You can prove it to yourself.

Everybody has two parents. That means, since each parent had two parents of their own, that we all have four grandparents. Then, since each grandparent had to have two parents, everyone has eight great-grandparents, and sixteen great-great-grandparents and thirty-two great-great-great-grandparents and so on.

You can go back any number of generations and work out the number of ancestors you must have had that same number of generations ago. All you have to do is multiply two by itself that number of times.

Suppose we go back ten centuries, that is to Anglo-Saxon times in England, just before the Norman Conquest, and work out how many ancestors you must have had alive at that time.

105

If we allow four generations per century, that's about forty generations ago.

Two multiplied by itself forty times comes to more than a thousand trillion. Yet the total population of the world at that time was only around three hundred million. Even today the population is seven billion, yet we have just worked out that a thousand years ago your ancestors alone were more than 150 times as numerous. And we've so far dealt only with your ancestors. What about my ancestors, and the Queen's and the president's? What about the ancestors of every one of those seven billion people alive today? Does each one of those seven billion people have their own thousand trillion ancestors?

To make matters worse, we've so far only gone back ten centuries. Suppose we go back to the time of Julius Caesar: that's about eighty generations. Two multiplied by itself eighty times comes to more than a thousand trillion trillion. That's more than a billion people packed into every square yard of the Earth's land area. They'd be standing on top of each other, hundreds of millions deep!

Obviously we must have done our sums wrong. Were we wrong to say that everybody has two parents? No, that is definitely right. So, does it follow that everyone has four grandparents? Well, sort of yes, but not four *separate* grandparents. And that is exactly the point. First cousins sometimes marry. Their children have four grandparents, but instead of eight great-grandparents they only have six (because two great-grandparents are shared).

Cousin marriage cuts down the number of ancestors in our calculation. First-cousin marriages are not particularly common. But the same idea of cutting down the number of ancestors applies to marriages between distant cousins. And that is the answer to the riddle of the very big numbers that we calculated: we are all cousins. The real population of the world at the time of Julius Caesar was only a few million, and all of us, all seven billion of us, are descended from them. We are indeed all related. Every marriage is between more or less distant cousins, who already share lots and lots of ancestors before they have children of their own.

By the same kind of argument, we are distant cousins not only of all human beings but of all animals and plants. You are a cousin of my dog and of the lettuce you had for lunch, and of the next bird that you see fly past the window. You and I share ancestors with all of them. But that is another story.

HOW DO THEY KNOW ALL SNOWFLAKES ARE DIFFERENT?

Justin Pollard

historian

The first person to realise that every snowflake seemed to be different was a man called Wilson Bentley who was born in 1865. He grew up in the state of Vermont, in the USA, which has very cold, snowy winters. In fact, the USA has the highest annual snowfall of anywhere on earth, including Antarctica. He also lived in a very cold farmhouse. So cold that he found he could catch snowflakes on a blackboard and bring them indoors to look at without them melting.

Now, Wilson's mum had an old microscope, and one day when he was fifteen years old he decided to look at the snowflakes through it and what he saw amazed him. Each one was a beautiful six-sided shape, but each one was different.

Wilson Bentley wanted everyone to see how beautiful snowflakes were but even in his cold house they would eventually melt. Then he had an idea. Wilson persuaded his dad to give him a hundred dollars (a lot of money in those days, nearly £1,500 today) to buy a camera and a special attachment so he could take photographs through the microscope. Not many people knew how to do this at that time and in 1885 he became the first person ever to take a photo of a snowflake this way.

He kept taking these photos all his life and became known as 'Snowflake' Bentley. In the end he took 5,381 photographs of snowflakes and each one was different. In the summer, when there wasn't any snow, he photographed the smiles of pretty girls. He died in 1931 after catching a chill whilst out in a blizzard on another snowflake-collecting expedition.

But was he right that every single snowflake was different?

Each snowflake starts out as a tiny ice crystal in a cloud, which grows as it swirls and falls down to Earth. Its shape depends on lots of things, including how cold and damp the air is at exactly the place where the flake is at each moment of its life. So the chances of any two flakes swirling and falling in exactly the same way are very tiny.

But then a lot of snowflakes have fallen in the history of the world. In fact, there are a million snowflakes in just one litre of snow and a nonillion snowflakes could have fallen in the whole history of the world. That's a huge number. To give you an idea how huge, if you covered the whole world with a nonillion £5 notes the pile would be 55,620 kilometres thick right the way round.

So could two of those be exactly the same? The truth is we can never say for sure as no one can have looked at all of them. But mathematicians have estimated that out of all those nonillion flakes only two might have looked the same under Snowflake Bentley's microscope, and even then, if you looked at them under a really big microscope, you would see tiny differences.

WHY DOES TIME GO SLOWLY WHEN YOU WANT IT TO GO FAST?

Claudia Hammond

psychologist and radio presenter

The problem with time is that it warps and not always in the way you'd like it to. The clock says one thing, but your mind says another. If I told you to close your eyes now and guess, without counting, when two minutes had passed, you'd soon be bored. The time would feel long. But if you were watching *Doctor Who* the same two minutes would go by in a flash.

Have you ever had that feeling that a lesson is about to finish, only to glance at the clock and realise you're not even halfway through yet? This is particularly likely to happen if you're bored and want time to go fast. When you feel bored you start paying attention to time itself. You notice every achingly long minute. But when you're playing your favourite game the opposite happens. You're so absorbed that the last thing you do is to focus on time itself. You're having far too much fun for that. When you're enjoying yourself time speeds up. Think of that last hour before bedtime. Time just seems to disappear.

The reason time goes slowly, even though you're willing it

to go fast, lies in the way the brain counts time. No one knows exactly how it's done, because although we have eyes for seeing and ears for hearing, there's no special part of the body just used for measuring time. Yet we're surprisingly good at guessing when a minute has passed. This is something you could try for yourself at home. Get someone else to test you, but no cheating by counting one crocodile, two crocodile!

One theory of how the brain keeps time is that it counts its own pulses, the pulses it's using to do other jobs. Our brains are very active, even when we're bored and think we're not doing anything. Scientists think that when we're bored and start paying attention to time, those pulses speed up. Then the mind counts up those pulses, and we think that more time has passed than really has. In other words, the lesson you don't like is still going on. Time has slowed down, even though you wish it would go fast.

Our minds do weird things with time. When you have a really boring day doing nothing, when you're ill, for example, time goes slowly. But when you look back afterwards on that week when you were ill, it feels as though it went by fast. The reason is that you didn't do anything new so the week won't occupy much space in your memory, making it seem short when you remember it. Time is weird and we never quite get used to it.

WHO FIRST MADE METAL THINGS?

Neil Oliver

archaeologist

Long before there were metal tools, people made a lot of the things they needed from all sorts of stones. After hundreds of thousands or even millions of years spent looking out for useful rocks and pebbles, humans became very expert at spotting different kinds of stone.

So along the way, some of the more inquisitive people would have noticed that some rocks glittered or shone when sunlight hit them. Perhaps they spotted them glowing under the water in shallow river beds, or as sparkling stripes on cliff faces and in boulders. Some of the shiny pebbles in the rivers would have been little nuggets of gold, and a bit of experimentation would have revealed this new stuff could be hammered into different shapes between two hard stones.

The first gold objects might have been made many, many thousands of years ago. It would not be quite right to call them 'tools'. Those early gold objects might rather have been valued as a kind of jewellery, or as good-luck charms.

Copper is another metal that is found naturally in lumps. It can be shaped a bit like gold, while it is still cold, but it's even easier to shape if you warm it up over a really hot fire. From time to time the odd lump of copper may have accidentally dropped into a cooking fire. And it would only have taken an observant person to notice how heat made the copper soft, kind of like butter.

This is where it gets really interesting: as well as turning up in lumps, copper also appears as bright blue or green stripes within rocks. Such rocks would have been attractive and eye-catching, and likely to be picked up and carried home by people.

It's not hard to imagine how such rocks might have found their way into hearths, or pottery-making kilns – either by accident or as an experiment. If the fire was hot enough, then liquid copper would sometimes have leaked out of the strange blue-green rocks. Think how exciting it would have been to see that happen for the first time, and how unforgettable!

Since some people have always been observant and curious, it makes sense to imagine this discovery being made many times, in all sorts of different places, during the thousands and thousands of years of our history. We know for certain that people living at the eastern end of the Mediterranean Sea – more or less the land we call Turkey today – were making copper tools seven thousand or maybe even eight thousand years ago. But other people learned how to make metal in other places, too. They were busy doing it in what is now modern Bulgaria by six thousand years ago, and by at least five thousand years ago people in what is now Pakistan were making metal too.

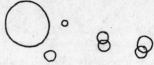

HOW DO THE BUBBLES GET INTO FIZZY DRINKS?

Steve Mould

science broadcaster

You know how you can dissolve things like sugar in water? What happens is, all the little bits that make up the sugar granules separate from each other and spread out. These little bits are called molecules and they're so small you can't see them. That's why the granules seem to just disappear!

Well, you can do the same thing with bubbles of gas. But to get bubbles to dissolve you have to squeeze them really hard. That is to say, you have to apply loads of pressure. Which is why when you open a fizzy drink you hear a hissing sound. That's the pressure being released.

And what happens when you release the pressure? All those little dissolved molecules come back together and form bubbles again. If you drink really quickly after opening a can, loads of the bubbles will grow in your stomach and you can do massive burps.

117

WHY IS THE SKY BLUE?

Simon Ings

author of books on science

Guess what? The sky is not blue. At least, there's no blue *stuff*, no blue pigment, in the sky. It's a trick of the eye. Up there, and all around us, are gases of different kinds, such as oxygen, nitrogen and carbon dioxide. There's also dust, water vapour, spores and even tiny airborne animals.

When sunlight hits something, it gets reflected. Big objects, such as the Moon, reflect the light very well. Moondust is dark, but so reflective that the Moon shines brightly in the night sky. But a tiny gas molecule is too small to act as a mirror. Instead, it absorbs light, and then sends that light bouncing back out again in a random direction. In other words, every molecule in the air is a tiny, flickering light source.

Imagine for a moment that light was sound. Sunlight is not just one note of a certain pitch played on one instrument; it is a vast orchestra playing every imaginable pitch at every imaginable volume! We see just some of this music. Our eyes perceive different pitches of light as colours: violet, blue, green, yellow, orange, red and purple.

Air molecules absorb blue light very easily, and they send it bouncing away just as easily. This is why blue light is scattered all around the sky, and why it reaches our eyes from all directions. Everywhere we look, we are bombarded with blue light. That is why the whole sky looks blue.

The other colours aren't nearly so easily scattered by the Earth's atmosphere, and they come to us in a more or less straight line. DON'T look directly at the Sun, because if you do, every colour there (aside from a little sky blue) will be hitting the back of your eye at the same time. That much light really can damage your eyes.

If Mars had more gas in its atmosphere, it too would have a blue sky. As it is, there isn't enough gas for this scattering effect to work. If you could stand on the surface of Mars and look up, you'd see the sky there is the white of raw sunlight, tinted beige by dust.

Towards the Earth's poles, the Sun sits low in the sky and sunlight has more atmosphere to pass through before it reaches the ground. Here the sky is *especially* blue.

HOW DO SPORTSPEOPLE CONCENTRATE WHEN THE CROWD IS NOISY?

Colin Montgomerie

golfer

I'm a golfer, and golf is a different game from other sports. It is an individual sport and a lot of the challenge is mental. In an individual tournament you have to focus so hard on what you are doing that you don't hear too much – and golf crowds are generally very knowledgeable and respectful. If you've got a big crowd following you, and it's noisy, it usually means you're playing well, so you shouldn't complain.

In team events like the Ryder Cup the crowd can be huge and sometimes as loud as a football crowd, with lots of chanting and singing, and huge roars. When it's a home crowd it can encourage you to play well and it gets the adrenalin going. There's nothing better than hearing the crowd chant your name. When it's an away crowd it can be very nerve-racking and some crowds can be quite unkind so that's when it gets harder to concentrate. You can't take the shouting personally, you just have to try to block it out, or to feed off it and let it inspire you to play even better.

The best way to concentrate is to forget about everything that's going on around you. Just think about your next shot. You just have to learn to trust the crowd and hope that no one decides to shout out just as you take the club back on your backswing, or just before you hit a putt.

I think the more experience you have of big crowds the easier it is to concentrate. You get used to the cheers and the noise they make, and remember that playing in that environment is the reason we put in all that practice over the years. If we want to be the best we should want crowds to watch us play and encourage us – it means we are playing well and hopefully have a chance of winning a tournament.

DO MONKEYS AND CHICKENS HAVE ANYTHING IN COMMON?

Dr Yan Wong

evolutionary biologist and science broadcaster

More than you can possibly imagine. For a start, think about what they look like on the outside. They both have a front end (a head with two eyes, a mouth, a brain, and so on), a back end (anus and tail), two legs (complete with knees and toes), and two 'arms'. Admittedly, the need for flying means a chicken's arms look a bit different, so we give them a special name: wings. But you'll find the same basic bones in the wing of your roast chicken dinner as in your own arm or that of a monkey.

Biologists call this deep similarity 'homology', and it's even more evident when you look under the skin. Monkeys and chickens have the same organs (lungs, heart, liver, kidneys), doing the same jobs. Under a microscope the animals have even more in common. Their bodies are built of the same basic cells, working in almost identical ways. Zoom in further still, to examine the tiny molecules that control life's chemical reactions, and you'll find the vast majority of these look almost identical.

There's a good reason why monkeys and chickens have so much in common. They are originally descended from the same animal: a lizard-like creature that lived about three hundred million years ago. From this common ancestor, they've inherited the same DNA – the same set of 'building instructions'. Monkeys and chickens look a little different only because these instructions have changed very slightly since then.

In fact, all life is related. Animals, including monkeys, chickens, and ourselves, share a common ancestor with, say, a tree. We don't think of ourselves as having much in common with trees, because our common ancestor lived over a thousand million years ago. But the intimate details of our biology reveal it to anyone who looks carefully enough.

HOW DID WE FIRST LEARN TO WRITE?

John Man

author of books about writing

Once upon a very long time ago, before writing, people had to remember what they said to each other because there was no way of recording conversations. That worked when life was simple. If you just wanted to exchange your chicken for a basket of your neighbour's apples, for example. Or if you wanted a priest to say a prayer to the gods for you, in exchange for one of your chickens.

But what if you wanted the apples or prayers right now and your neighbour or your priest wanted the chicken tomorrow, or next week, or next spring? What if, when the time came, your neighbour or priest said, 'But you said *two* chickens!' and you couldn't quite remember what you had said. There must have been a lot of arguments about who said what, when and to whom.

To find out how writing started you have to look back ten thousand years, to today's Iraq, then known as Mesopotamia. This was a hot area, with two great rivers, the Tigris and the

Euphrates. Mesopotamia means 'between two rivers'. Big rivers are good for food and transport. They give water for crops, which can be carried in boats, and they provide drinking water for cities.

In a big, complicated, rich country like Mesopotamia, people needed to make records of what was happening, especially the priests. Beneath their feet, in the earth, which was often flooded, they found what they needed – clay. It was easy to make little, soft clay balls, and then use a reed as a sort of wooden pen to make signs that meant things like 'two chickens' or 'seven sheep'. You can do the same thing today: all you need is some mud and a twig.

Because the pens had three-cornered nibs, they made marks that were triangular, so their writing is called cuneiform, which means 'shaped like a triangle'. Then they baked the clay in an oven, and kept the hard balls in their offices to avoid arguments in months and years to come.

Later, scribes learned to make lots of different signs to record any word in their language. They could record many different things: accounts of wars, lists of kings and officials, and the stories that parents told their children. Scientists have now dug up tens of thousands of these clay tablets and scholars know what the signs mean.

Time passed, and two other great rivers made their regions big and rich. One was the river Nile in Egypt. About five thousand years ago, priests in Egypt made up different signs. They wrote on temple walls and on a sort of paper made from reeds,

126

recording stories of gods and kings in picture-writing called hieroglyphs, which means 'sacred writing'.

Then, two thousand years later, in China, people built cities on their great river, the Yangtse. Here priests did something really strange. They used to heat up the shells of turtles in fire to crack them, and then used the cracks as some people use tea leaves today, to tell the future, scratching their judgements beside the cracks. This was the foundation of all Chinese writing. Hundreds of these cracked turtle-shells have been found, and they show that a few of the signs from three thousand years ago survive today.

WHY DO SCIENTISTS LOOK AT GERMS, AND WHY CAN'T I SEE THEM?

Joanne Manaster

biologist and science educator

When we talk about germs, we are often talking about bacteria and viruses that can make us sick. It is amazing that very tiny organisms, too small for just our eyes to see even through a magnifying glass, can make us feel so horrible at times!

Our eyes are only able to see clearly objects that are two hundred micrometers or more in size, and a single one of your hairs is about that thick. Most bacteria are about one micrometer in size, meaning two hundred will fit across the width of the human hair.

To see bacteria, we can use a light microscope, and we can see that some germs are little balls, some are tiny rods and others look like a spiral corkscrew. Sometimes they stand alone and others like to form chains or clump with each other. With special coloured stains we can see that some bacteria are different from others. Some look purple and some look pink.

When scientists were able to use even more powerful microscopes, they learned that the walls or boundaries of bacteria could be different. These differences gave clues to the scientists about how bacteria behave to make people sick. Some bacteria are able to make people very, very sick because of how they are designed. Bacteria might have 'tails' to help them swim and infect cells easier. Some have tiny hairs all over their membrane to help them stick to cells, like the ones in your throat, and some even use a slimy, wet coat to help them live longer in dry conditions. Once scientists knew more about how the bacteria were constructed, they developed better drugs that could break apart the different types of bacteria and help our bodies fight the sickness.

If you go to the doctor with a skin infection, you might be given an antibiotic. When you go for another kind of illness, like a sore throat or a very bad stomach ache, you might be given a different type of antibiotic because the doctor usually knows by your symptoms which bacteria are infecting you, and knows which drug will destroy the bacteria better. These drugs are designed to work on destroying the structure of bacteria and the way they work, based on what we know by looking very closely at them.

Sometimes you go to the doctor and don't get any antibiotics, and that could be because you have an infection that is caused by a virus. Viruses are tinier than bacteria, and they look and behave differently, so the drugs meant for bacteria will not help at all.

DO ANY PEOPLE EAT POLAR BEARS OR LIONS?

Benedict Allen

explorer

No, it would be far too much trouble. When it comes to look-ing for a tasty morsel to eat, people find it easier to catch ani-mals that do not have gigantic teeth or dreadful claws. It is true that both lions and polar bears are very good at finding people. So although they're much less bothersome to get hold of – because you don't have to run after them – the problem is that you are half eaten before you can think about how to cook your polar bear or lion.

When I was in Outer Mongolia, which is at the other end of the world, I stayed with very kind people called the Tsaatan, who lived in tents made of reindeer skin and wandered with their herds of deer through the snowy forests. One evening a man came into the tent which I shared with a family. He looked very tired. He said that he had just been chased by a bear, and it wanted to eat him because he was dressed in clothes made from lovely warm reindeer skins. The bear thought he was a reindeer, because he smelt like one! However much he shouted, the bear ran after him. So it was all a bit of a muddle.

And when the bear saw at last that he wasn't actually chasing a reindeer he decided he might as well eat the man anyway. In the end the man had to scare the bear away with his penknife. The blade was not as good as the bear's claws, so it took a long time. So you can understand why the man was so tired. I gave him a cup of tea, and a kind lady repaired his clothes, which were ripped.

And lions are the same, you see. It's just not so easy to eat them. Once, I walked through the Namib Desert, which is a dry place in Africa, with my three camels. My favourite camel was called Nelson, and if there was one thing he didn't like – apart from giraffes, which were taller than him – it was lions. He didn't like the way they wanted to creep up on us. I felt the same. I did not like it one little bit. Both Nelson and I wanted to go home.

The lions used to circle about us at night. I was sure they were trying to decide which one of us was best to eat. I was sure they would choose me, because camels can run fast, and can jump up and down on big things that creep up on them. But people can't.

So, for that reason, although people do eat monkeys and snakes and bats and even spiders, they like to stay away from polar bears and lions. And they hope polar bears and lions stay away from them.

WHY DOES THE MOON CHANGE SHAPE?

Professor Chris Riley

science writer and broadcaster

Everything, and I mean everything, in the universe is on the move! And the Earth and the Moon are no exception. Right now, as you read this book, you and the book, and your house, your street, your neighbours and everyone you know are hurtling through space at over twenty-seven kilometres a second, as the Earth travels around the Sun.

And if you can see the Moon out of your window, take a good long look and remember that it's also hurtling around the Earth at more than a kilometre each second. I know it's hard to believe, as you can't see it moving at all. But that's because it's really a very long way away: about 385,000 kilometres away, which is the same distance as travelling ten times around the Earth.

At this distance from the Earth, the Moon takes almost a month to travel once around us. You've probably noticed that during this time it changes appearance, growing from a thin sliver or crescent into a whole circle, then back to a thin crescent again before completely disappearing for a day or so. How can we explain such dramatic changes? Any ideas? Time for an experiment.

You'll need a dark room to simulate space, a lamp (for the Sun) and an apple (for the Moon). You will play the part of the Earth!

Switch on the Sun (your lamp) at one end of the room and turn all the other lights off. Stand up and hold the apple out at arm's length towards the light.

With all the light falling on the opposite side of the apple, the side of the apple you can see should be in darkness. Now turn on the spot to your left through one eighth of a whole circle – still holding the Moon (your apple) at arm's length. What does it look like now? You should be able to see a thin sliver of your 'Moon' illuminated on the right.

Turn another eighth of a circle to your left. The Moon will now be half illuminated by the Sun (your lamp). Keeping it at arm's length, make another quarter-turn to your left. Now the 'Sun' should be behind you, and as long as you aren't casting a shadow on the 'Moon' you will see that the side of the apple facing you is now fully illuminated, like a full Moon. Continue turning to your left, holding the 'Moon' out at arm's length, and you'll see that the illuminated part starts to shrink again, first back to a half-circle and then a crescent, before disappearing as you bring it back to where you started.

What you've just simulated is exactly what happens to the Moon as it travels around the Earth at one kilometre a second! Your experiment also proves that the Moon isn't a flat disc, as it sometimes seems to be in our night skies, but a spherical world like the Earth, illuminated from a single direction by the Sun.

DO NUMBERS GO ON FOREVER?

Marcus du Sautoy

mathematician

Here's one of my favourite mathematical jokes to help answer this question:

A maths teacher asks the class: 'What's the biggest number?'

One of the kids quickly puts up his hand. 'A trillion,' he announces.

'What about a trillion and one?' the teacher responds.

'Well, I was close,' the child replies triumphantly.

The reason this is funny (of course it always kills a joke to have to explain why it's funny) is that the child thinks the teacher's answer of 'a trillion and one' is actually the biggest number that exists. In fact, the teacher is giving an answer to the question 'Do numbers go on forever?'

If numbers didn't go on forever, it would mean there had to be a biggest number. But if there was a biggest number, I could play the same trick as the teacher. I could add one to that number, and now I've got an even bigger number.

The numbers never run out. They do go on forever.

WHERE DID THE FIRST SEED COME FROM?

Dr Karen James

biologist

When you think of the word 'plant' you probably picture a flower, a tree or maybe a field of grass. All of these plants grow from – and produce – seeds. But there are other kinds of plant that don't come from seeds at all. Ferns and mosses don't have seeds or flowers but reproduce using spores. Spores are kind of like seeds, but there are some important differences (more about that later). There are still other plants called algae that live in the water and don't produce spores *or* seeds, but reproduce in other ways.

Around 350 million years ago, scrubby moss forests had given way to more impressive forests of tree-like ferns. Insects and spider-like creatures scuttled about, taking advantage of the food and shelter provided by these plants. In the water, the fins of some fish were evolving into legs that enabled them to walk on land. These became the amphibians – the ancestors of frogs, toads and newts.

It was during this time that the spores of some fern-like plants evolved to become larger, with a starchy food source inside and a waterproof coat. These were the first seeds. Their food sources gave young plants a head start in difficult environments and their waterproof coats helped them survive in dry, inhospitable places – places where a spore would have no chance.

When the naturalist Charles Darwin was writing his famous book *On the Origin of Species*, he did experiments at his home at Down House in Kent to show just how long different kinds of seeds could survive in sea water. (Most seeds like fresh water, so sea water counts as an inhospitable place.) From this, he did some maths to show how far the seeds might be able to travel across an ocean. This was important because in Darwin's time it wasn't understood how plants could be living on, say, distant islands, unless they had been specially created there. Darwin showed that they could have travelled there as seeds, across the ocean and, once there, evolved into new species.

Seeds' waterproof coats help them to survive not only in dry places and in the ocean, but sometimes for a very long time. In 2005, scientists in Israel successfully germinated a two-thousand-year-old seed!

All of these benefits of seeds are what helped the early seed plants to be successful so many millions of years ago. So the next time you walk through a meadow, put on a cotton shirt or eat a bowl of oats, remember the ancestors of those plants. And how, by storing their energy and 'wearing' a waterproof coat, they evolved into the hundreds of thousands of beautiful – and useful – kinds of plants with which we share the Earth today.

WHY WAS GUY FAWKES SO NAUGHTY?

Philippa Gregory

author of historical novels

Guy Fawkes was *extremely* naughty since he planned to blow up the King of England, which was a bad thing to do, even in 1605. But he would have said that it was the only way to save England for the Catholic religion. Guy (who was also called Guido) was a Catholic – a Christian who believes that when the priest offers the bread and wine at Mass, they really and actually become the body and blood of Jesus. All Christians in Europe had believed this until reformers started to think that although the bread and wine should be called the 'body and blood' they didn't actually change.

Like many religious arguments this might have been something that people just talked about, but Henry VIII and then his daughter Queen Elizabeth I made this new reformed church the only religion allowed in England. So the people who still believed in the Catholic view, and who wanted to obey the Pope, found themselves named as criminals, facing harsh punishments.

James I became king after Queen Elizabeth, and Guy Fawkes felt that he had to stop the new King from forcing the reformed religion on England. He thought that the best way to do this would be to blow him up, along with lots of other important people when they met for the opening of parliament.

He and four other men collected thirty-six barrels of gunpowder, but their plot was discovered and Guy was caught guarding the gunpowder. He was taken to the Tower of London and tortured till he confessed.

He was condemned to a terrible death of 'hanging, drawing and quartering'. This meant hanging by the neck until the prisoner was nearly dead, then being taken down and cut in the belly to take his insides out, which were then burned as he watched, until he died. Then they cut up the body and sent it to the four quarters of the kingdom, so that everyone knew a traitor would be terribly punished.

It was clear – even to King James who he had tried to kill – that Guy had only done what he thought God would want. But there was no mercy for him. Naughty to the end, he managed to escape the terrible pain of the full execution. When he was on the scaffold he managed to jump down, break his own neck and die quickly.

Many people were so pleased that the King was safe and England at peace that they lit bonfires. Later, the government ordered that people should celebrate on 5 November every year. This is Bonfire Night, when bonfires are burned with 'a guy': Guy Fawkes.

WHAT DO YOU HAVE TO DO TO GET INTO THE OLYMPIC GAMES?

Jessica Ennis

athlete

Make sure you train hard, look after yourself both physically and mentally and never let bad days get on top of you as a good one is around the corner.

WHO WAS THE FIRST ARTIST?

Michael Wood

historian

It's a great question and you've asked it at a time when we've just made an amazing discovery. A prehistoric paint kit was recently found in the Blombos cave by the seashore in South Africa. It is probably more than ninety thousand years old! They found cut seashells, containing red and yellow paint colours, along with grinding stones and bone spatulas for mixing the paint. We think the people who made this paint kit would have used their fingers to paint on their bodies and cave walls.

Human beings are above all creative beings and we must have been painters, carvers, shapers and whittlers before we could even speak language. But who were the earliest artists? Prehistoric paintings have been found all over the world, many of them haunting images that show us the human imagination taking flight. Look at the mazy, geometric shapes of native Australian art, the mysterious cosmic patterns of India or the swirling hunt scenes in the caves of southern France, and you touch the mystery of artistic creation itself. These are messages to us from our ancestors, who felt the need to paint, to leave behind their responses to the world around them and to the universe itself.

Of course, we can never know who the earliest artists were. But artists they were. Take the ancient, tiny carving of a woman called the Hohle Fels Venus found in 2008. The figurine is only six centimetres high, carved from woolly mammoth tusk. But when you look at her you can see that she must have been carved by someone with incredible sensitivity. This is from forty thousand years ago, a period when art – and maybe music too – seems to have made a great leap forward.

And the earliest great art? There is so much to choose from, but my favourite early paintings are the Altamira cave paintings in Spain. These first fascinated me as a child and the images are still, literally, awesome. The animals are fantastic: bison in deep luminous oranges edged in black, their movements captured with amazing liveliness. When they were first found in the nineteenth century some people argued that they were modern fakes because prehistoric humans could not have had the skill or intelligence or vision to create such things. How wrong they were!

WHAT AM I MADE OF?

Professor Lawrence Krauss

particle physicist and cosmologist

Stardust. Well, sort of.

Everything in your body, and everything you can see around you, is made up of tiny objects called atoms. Atoms come in different types called elements. Hydrogen, oxygen and carbon are three of the most important elements in your body.

Water actually makes up most of your body's cells. You are almost ninety per cent water. Each water molecule contains two hydrogen atoms, which are light, and one heavier atom, oxygen.

It turns out that atoms are actually made of even smaller objects, particles called protons, neutrons and electrons. Protons and neutrons in turn are made up of even smaller objects called quarks. As far as we know, electrons and quarks aren't made up of anything smaller.

So, why are you made of stardust?

Our universe began in a big explosion, called the Big Bang, over thirteen billion years ago. But in that explosion only the

145

very lightest elements were formed out of protons, neutrons and electrons. The place where the heavier elements were formed, like the oxygen and carbon that are so important in our bodies, was in the fiery furnaces in the centre of stars, where the temperature can exceed hundreds of millions of degrees.

How did those elements get into our bodies? The only way they could have got there, to make up all the material on our Earth, is if some of those stars exploded a long time ago, spewing all the elements from their cores into space. Then, about four and a half billion years ago, in our part of our galaxy, the material in space began to collapse. This is how the Sun was formed, and the solar system around it, as well as the material that forms all life on earth.

So, most of the atoms that now make up your body were created inside stars! The atoms in your left hand might have come from a different star from those in your right hand. You are really a child of the stars.

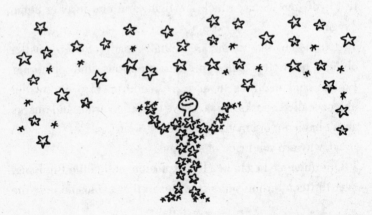

WHY DO PENGUINS LIVE AT THE SOUTH POLE BUT NOT THE NORTH?

Vanessa Berlowitz

TV documentary producer

Although penguins don't get quite as far south as the South Pole, they do tend to live in the freezing cold seas that surround Antarctica in the southern part of the planet.

They do well here because they come with some of the best cold-weather gear I've ever seen! They have feathers on the outside that lock together like tiles on a roof to make a sealed, waterproof coat which they wear over layers of fluffy down. Being nice and fat also helps keep them warm. They would have a really hard time relocating up to the Arctic, in the north of the planet, because they would have to swim through the warm seas around the equator to get there. Imagine how uncomfortable that would be for them. A bit like you or me having to run around wearing a full ski-suit on a hot, sunny day.

When we were filming *Frozen Planet*, I was surprised to discover that penguins have to work really hard to keep cool in

147

summer, even in the Antarctic where it never gets much hotter than a mild English winter's day. The whole team laughed when we watched the first footage that came back of king penguins flopping down in cold, wet sand to cool off their bellies and lose heat from their naked pink feet. Their chicks were even funnier going for mud baths to keep cool and coming out looking like they had been dipped in melted chocolate!

If penguins could travel to the north without overheating on the way, they would bump into a group of black-and-white birds called the auks that look a lot like them. Except that auks can fly and penguins can't. And that's another good reason for penguins not to move north to the Arctic – they wouldn't be able to fly away from the polar bears and Arctic foxes that prowl the colonies of nesting birds there in summer.

In the Antarctic, penguins don't have to worry about being hunted when they nest. There are no predators that live on the land, as none of their ancestors were able to swim across the rough, cold seas to get there. Some of the group of birds that penguins originally came from lost their ability to fly – because they didn't need to escape from predators. Penguins' wings are stubby and short. They use them like flippers, to propel themselves underwater.

I was lucky enough to watch penguins swimming, from overhead, while I was filming them from a helicopter in Antarctica. It was then that I realised that penguins could fly in a way, at least in the sea. It was one of the most beautiful things I have seen. Like watching an underwater ballet.

Everyone thinks of penguins on land with their silly, waddling walk. But it's only when you see how graceful they are at swimming in these cold, southern seas that you realise this is their natural home.

HOW DOES AN AEROPLANE FLY?

David Rooney

transport curator at the Science Museum, London

When you first fly in an aeroplane, it doesn't seem possible that such a big, heavy thing full of people and luggage could fly. Heavy things like being on the ground. And really heavy things like aeroplanes *really* like being on the ground.

But there's no need to worry. Just take a look at birds flying around. They're pretty heavy, but they manage to stay in the air. And they do so using a very neat trick of nature.

You'll have seen that aeroplanes have long bits that stick out at the sides, called wings. If you've been in an aeroplane you'll also know that just before you take off, the pilot drives all the way to the end of a long road called a runway, then turns round and starts driving forward along the runway very fast indeed (this is the most exciting bit of the journey for me).

Now, here's where nature's flying trick kicks in. As the aeroplane moves forward, the air it moves through flows over the wings. When you run forward really fast you can feel a breeze on your face – it's the same idea.

By the way, the plane's wings are shaped sort of flat with a bit of a curve. This shape means that the air changes direction in order to flow over and under the wings. And when air changes direction in this way, it pushes the wing upwards. I can't really explain *why*. It just does.

So, as long as the aeroplane is moving forward, the air pushes the wings upwards and the aeroplane flies.

By now you might be wondering how the aeroplane moves forward so fast. This is caused by engines. Most aeroplanes these days have either two or four engines, and most use a type called a jet engine. (This is why sometimes aeroplanes are called jets.)

Jet engines burn a liquid fuel such as kerosene. And when it burns, it makes a jet of very hot gas stream out behind it. This jet of hot gas pushes the aeroplane forward, or spins a fan that does the same thing. You'll probably have discovered that jet engines are *incredibly* noisy. It's all the burning that's going on.

There are lots of other bits and pieces aeroplanes need to fly. For instance, how do they steer? And how do they slow down? There are flaps on the wings and at the aircraft's tail. The pilot makes these flaps move up or down, and together they can make the aeroplane speed up, slow down, go up, go down or turn left or right.

Those are the basics – how planes move forward and lift off and steer and land. They're pretty amazing things, really.

WHAT'S THE STRONGEST ANIMAL?

Steve Leonard

vet and wildlife TV presenter

Well, this is a toughie. Obviously, we can look at the animal that can lift the heaviest weight and this would probably be the elephant. Asian elephants have been recorded as lifting three hundred kilograms with their trunks, which is a good start. But if you wrap a leather rope around logs and get an elephant to bite onto the rope it can lift as much as five hundred kilograms, which is half the weight of a small car. This may seem like a lot but compared to the weight of the elephant it's only a small fraction. It's like me picking up nine bags of sugar, which I can easily do with one hand.

So maybe we should look at muscle power compared to body weight. The strongest people in the world can only lift weights twice their own weight. This is pretty impressive but nothing compared to some other animals. Male gorillas are very strong and can lift ten times their own weight, making them five times stronger than a human! But the truly strongest animals on earth for their size are the bugs. Leafcutter ants can lift pieces of leaves fifty times their own body weight. This is like me lifting a female Asian elephant into the air!

However, it gets better. A dung beetle can lift something 1,141 times its own weight, which is like me lifting six double-decker buses! Some microscopic creatures may be even stronger but getting them to lift stuff is very difficult.

WHO NAMED ALL THE CITIES?

Mark Forsyth

blogger and author

Cities usually get their names from people who live round about and are just describing what's there. Sometimes you can see that and sometimes you can't.

Newcastle and Oxford are pretty easy. There was a new castle so it was called Newcastle. There was a ford in a river where oxen could cross, so it was called Oxford.

But sometimes you can't tell any more. And that's because languages change. Think of all the words that you use in the playground that your grandparents don't understand. And think of all the funny phrases that they use that sound a bit old-fashioned. Well, that's nothing new. Your grandparents had the same thing when they were little and so did theirs, and theirs, and theirs, and theirs all the way back for centuries.

So your great-great-great-great-great-great-grandmother might have said that something muddy was 'liver'. And when she heard that somebody came from the city of Liverpool she would have known that that meant it was the city by the muddy river.

And your great-great-great-great-great-great-great-great-great-great-great-great-grandfather might have heard the name Birmingham and said, 'Oh right, that's where the Birm family's farm is.'

It would take about a hundred greats before you got to anyone who knew what London meant, but it was probably just a place by a river so deep that you couldn't wade across it.

Occasionally, towns get their names in other ways. Alexander the Great just built himself a city and called it Alexandria. And Khartoum, the capital of Sudan, means 'The End of an Elephant's Trunk'. But I've no idea why.

WHY IS WATER WET?

Roger Highfield

*director of external affairs at the Science
Museum Group*

One answer is to say that when you touch a puddle of water it
feels wet because your fingertips tell your brain that the sensa-
tion is 'wet'.

Nerve impulses are sending messages from your skin to
your brain all the time about the world around you. We call
this your sense of touch. Your sense of touch also tells you
when something is dry, hot or cold, rough or smooth. Water
feels wet, which means that water is a liquid.

But water is only a liquid between zero degrees celsius and a
hundred degrees celsius. At zero degrees or cooler it is solid ice.
If you take ice cubes from the freezer and put them in a drink
at room temperature, the ice warms and starts to melt. Melting
makes them liquid again. And when water in a kettle heats up
over a hundred degrees it becomes a gas called water vapour,
which is invisible to our eyes. (When you see steam come out of
the kettle it is actually tiny drops of liquid water that form as the
hot water vapour hits the cooler air around the kettle.)

157

If you had a super-microscope, you would see that water is made up of little particles called molecules. Each molecule is itself made up of smaller particles called atoms. You can think of these like Lego blocks, which make up the molecules used to build all the stuff (chemicals) around you, and all the stuff in your body too.

Each water molecule consists of two atoms of hydrogen stuck to one atom of oxygen. Molecules stick to each other too but it turns out that water molecules 'glue' each other together with hydrogen in an unusual way. You can learn the details of this special glue when you are older. All you need to know for now is that these 'hydrogen bonds' hold together water molecules more tightly than other similar-sized molecules that don't have them. That makes water weird in all kinds of ways.

Here are some of the ways water is weird:

- Liquid water has a thin 'skin' on the surface. You can't see it but it's strong enough for insects to walk on. This skin means that liquid water sticks to our hands and clothes and feels wet. Some other liquids, like the metal mercury, don't feel wet at room temperature because they don't have this slight stickiness. If you poured liquid mercury onto your hand, it would just roll off like marbles. (Don't do this, though, because mercury is nasty stuff!)
- Water boils and melts at a much higher temperature than substances with similarly sized molecules.
- Most substances shrink when you cool them, but water

expands when it freezes. This is because the special hydrogen bonds hold its molecules further apart. So ice takes up more space than liquid water. That's also why ice cubes float.

- Clever experiments by Rich Saykally at Berkeley, California, and sums by David Clary, now at Oxford University, have shown that if you want to get wet, you need at least six molecules of water. If there are fewer, the molecules form films just one molecule thick. Add a sixth molecule to the group, and the cluster of molecules flips to a microscopic puddle, which feels wet to us.

WHAT WOULD I LOOK LIKE IF I DIDN'T HAVE A SKELETON?

Professor Joy S. Gaylinn Reidenberg

comparative anatomist

If you didn't have a skeleton, you might be able to stretch your arms like rubber bands, flatten yourself so you could slip under a door, or reshape yourself like a Shapeshifter from *Harry Potter*!

However, you'd have some definite drawbacks. It would be hard to hold those shapes against the force of gravity. Most of the time, you'd end up the same shape as whatever box or bowl you found yourself lying in – like water in a cup or gelatin in a mould. Without a container, you'd probably look like a big, wiggly plop of jelly that had spilled on the floor.

Your skeleton gives you form – an internal frame that helps you hold your shape. It provides a surface where muscles can attach, as well as joints that act like pulleys and levers. Without any hard parts against which you could pull your muscles, and without the mechanical advantage of joints, you'd be very weak and tired because you'd have to use more energy to move your arms and legs.

161

If you chose to live in water, you'd be nearly weightless, and therefore you wouldn't be so tired when you tried to move. You'd probably look a lot like a jellyfish, squid or octopus. I once dissected a giant squid and got to see just how unusual its body was. These animals do not have bones, but they do have amazing flexibility because they can bend pretty much anywhere – not just at joints like we do. Imagine being able to coil your arm into a spiral!

It reminded me of my dissection of an elephant's trunk, which could bend in several directions by muscle action alone, without any bones! The arm of a squid works in a similar way. It bends when muscles pull back on one side only, it shortens when all the muscles pull back at once, and it lengthens when an outer ring of muscles clamps down like a fist around the core. This last action pushes the fluid inside towards the tip – like your hand squeezing a tube of toothpaste – and shoots the arm forward.

I once had an exciting encounter with a live giant Pacific octopus while scuba-diving. I loved to see how it changed its shape: wrinkling its skin so it looked like rocks or sea-weed, flattening its arms so they were shaped like aeroplane wings, or coiling and uncoiling the arms under its body so they appeared to roll like wheels.

The most amazing moment was when it reached out its arms to touch me (right before it crawled over my face and completely covered the view from my dive mask with suc-tion cups!). This action of uncurling and extending its arms reminded me of party blowers!

162

ARE COWS POLLUTING THE AIR?

Tim Smit

chief executive of the Eden Project

Yes but . . . cows do good stuff too.

So, first things first, how do cows pollute the air? It's all to do with what they eat and how they eat it. Unlike you or me, cows have stomachs with four compartments. This allows them to eat grass, which is tough, chewy and takes ages to digest. They store the eaten grass in the first part of their stomach so they can bring it back up and chew it later, which helps break it down. That's why they look like they're chewing gum.

The second part of their stomach is full of useful bacteria, which break down the grass even more. This process produces a pongy gas called methane, which the cows breathe out. Humans sometimes produce methane too, often after eating too many baked beans, but with us it comes out the other end. Parp, excuse me!

Oh, just in case you were wondering, the third and fourth parts of the cow's stomach behave a bit like our (single) stomach. But that isn't much to do with the pollution story, so no more about that.

Back to the pongy, polluting gas. Methane is a greenhouse gas which, like carbon dioxide, forms a gassy blanket round the Earth that keeps the heat in, adding to climate change. Methane keeps the heat in more than carbon dioxide and comes from other places as well as cows' mouths and other animals' bottoms, including fossil fuels (coal and oil), marsh gas produced in wetland areas and gas from rice paddies. Livestock (cows, sheep and goats) produce about as much methane as the fossil fuel industry, less methane than marsh gas and more than is produced by growing rice.

Eating less meat means fewer cows and less methane, so that's one way to help reduce greenhouse gases. However, cows do good stuff, too. Some land isn't suited to growing crops for humans – like wheat for bread or beans – but can grow grass for animals to eat. Also, around a billion people worldwide depend on livestock to help make a living, including seventy per cent of the 880 million poor people in rural areas who live on less than a dollar a day. Grown-ups can check where the meat they buy comes from and whether it has been responsibly sourced.

There are lots of other things you can do to reduce greenhouse gas emissions, like saving energy by turning off lights, computers and TVs when you are not using them, helping grown-ups to use the car less, recycling things, sharing ideas with friends and family – and using your imagination to come up with new ideas.

Talking of new ideas, scientists in Australia have discovered that kangaroo gut bacteria produce less methane than cow gut bacteria. So they are trying to work out how to put these kangaroo bacteria into cows to make them more environmentally friendly.

HOW DO WRITERS THINK OF THEIR IDEAS?

Philip Pullman

author

I think if you asked ten different writers this question, you'd probably get ten different answers. In ancient times, poets used to believe in the Muses, who were goddess-like beings whose job it was to inspire them. There were nine Muses altogether, one for epic poetry, one for tragedy, one for dance, and so on, and poets or musicians would pray to the Muse or perhaps make a sacrifice to her, in the hope that she would give them some good ideas.

I don't think anyone these days believes in the Muses, but I understand why they used to. Ideas come mysteriously; you can't guarantee to get a good idea just by calling yourself a writer. They seem to come from somewhere out there in the darkness, for no particular reason.

But it does help to be prepared. When people ask me where I get my ideas from, I sometimes say, 'I don't know where they come from, but I know where they come *to*: they come to my desk, and if I'm not there they go away again.' In other words, whether you're really at your desk or whether you're anywhere else, you have to be prepared to recognise a good idea, and to do something about it.

When I was at school I used to find playing cricket a good time for ideas to come. The reason for that was that I was no good at batting or bowling, and I couldn't catch either, so I was generally sent to the furthest part of the field where I could just hang around in a state of half-dream and half-attention. That state of mind is ideal for helping ideas to arrive. I think I've lived in that state for most of my life, actually.

Some writers carry notebooks around with them to write down an idea as soon as it arrives. That might work for you. I've tried it from time to time, but I never found it really helpful because a *good* idea for a story would stick to my mind like one of those burrs that catch on your clothes when you're walking in the country. I couldn't get rid of it even if I wanted to.

And they can come from anywhere. Lots of ideas come from reading, and there's nothing wrong with being inspired by another writer; most of us started by being so thrilled by something we'd read that we wanted to imitate it. A lot of ideas come from just watching and listening to people.

But *having a good idea* is only the start. What you have to do then is make it into a story. Some people think that all they need in order to be a writer is inspiration. Not a bit of it! Plenty of people have good ideas, but very few of them actually go on and write a story. That's where the hard work starts.

But don't worry about that: if you work hard, and regularly,

168

and keep going even if you're not feeling good about it, the Muse will see you doing that and reward you with ideas. And one of the best feelings you'll ever have is getting a really good idea to solve that problem you've been fretting about for weeks. It really happens, which is why I still – sort of – believe in the Muses. At any rate, I treat them with great respect.

WHO INVENTED CHOCOLATE?

Joanne Harris

author

Chocolate as we know it, in bar form, was the invention of Mr Fry of London in 1847, but chocolate has been used for thousands of years. The Mayans and the Incas, in Central and South America, used a kind of chocolate drink in their religious ceremonies and the habit was brought over to Europe by the early explorers.

Christopher Columbus is said to have brought the first cocoa beans back to Europe in approximately 1503 but no one was sure what to do with them. A few years later, Spanish conquistador Hernán Cortés discovered the 'New World' and when he returned to Spain from Mexico in 1528, he loaded his galleons with cocoa beans and the equipment to make chocolate to drink.

It took over a hundred years, however, before the custom of drinking chocolate spread across Europe to England. After that, drinking chocolate became very fashionable among the wealthy, and was even once denounced by the Pope because it made people greedy!

WHY DO MEN GROW BEARDS AND NOT WOMEN?

Dr Christian Jessen

medical doctor and broadcaster

You could also ask, 'Why do men look different from women?' It all boils down to two rather clever hormones or chemicals, which really start to work in your body when you reach 'puberty' at around thirteen years old. These hormones are called oestrogen and testosterone and they are what make you start to look more like a grown-up after puberty, and also make you look either male or female.

The hormone called oestrogen is most active in girls. It helps breasts grow, as well as other female parts. It also makes the hair on girls' heads grow long and stops hair from growing on their faces.

In boys the hormone called testosterone is more active. It makes their voices get deeper and is why they grow taller and develop more muscles. It also makes hair grow on their faces and other parts of their bodies, but it slows down the growth of hair on their heads. This is why you may see men with big beards who are bald on top!

So the answer to your question about why men grow beards and not women is because men have more testosterone in their bodies than women.

Sometimes women can have medical problems where their bodies make too much testosterone – that's the male hormone. If they don't ask a doctor to put the balance right, guess what happens? They can start growing beards too.

IS SUGAR BAD FOR YOU?

Annabel Karmel

parenting author

We are all programmed from birth to like sweet things. Scientists believe this is because poisonous foods such as some berries are naturally bitter, so sweet tastes are associated with food that is safe.

Not all sugar is bad. There are natural sugars that you can find in food such as fruit. These forms of sugar haven't been played about with and are not bad for you as long as you don't eat too much.

But sugar is added to all sorts of manufactured food, especially into savoury food you wouldn't think had sugar added, from soups and sauces to pizza, crisps and ready meals. This means you can add much more sugar to your daily diet than you think.

A lot of breakfast cereals are also stuffed full of sugar, sometimes as much as thirty-five per cent. This is not a good way to start the day as it will not give you enough sustained energy to get you through the morning. There is a debate at the moment over whether these cereals should be moved to the biscuit aisle in shops! My rule of thumb is: if you look at the label of something and sugar is listed in the top three ingredients, put it back on the shelf.

There are a couple of reasons you get told not to eat too much sugar. For one, it is bad for your teeth. Have you ever done that experiment where you take a tooth that has fallen out and drop it into a glass of fizzy drink? (A 2p coin will also work if you have no teeth to hand.) See what happens to the tooth or coin even after just a few hours!

Sugar does the most damage to your teeth when you eat it frequently, so it's best to eat sugary foods as part of your meals rather than as snacks between meals.

Sugar is also bad for other parts of your body if you eat too much every day. Eating sugar can change your behaviour. When you eat sugar, it goes into your bloodstream giving you a burst of energy, and your body produces something called insulin, to deal with this sugar. This spike in energy doesn't last very long and you can feel very tired and wobbly once the sugar burst has finished. If you eat lots of sugar your blood sugar levels keep on going up and down and up and down. Your body doesn't need all this sugar, so it stores the extra, which can lead to more weight gain than is healthy.

HOW DID THEY BUILD THE PYRAMIDS IN EGYPT?

Dr Joyce Tyldesley

Egyptologist

The Ancient Egyptians did not have electricity or complicated machinery. Nor did they have a large workforce of slaves. Instead, they relied on people power. Their pyramids were built by many thousands of workers who travelled to the building sites from towns and villages all over Egypt. They camped at the site, worked hard for a few months, and then returned home for a rest as new workers arrived to take their place. They were supervised by a small team of expert builders, stonemasons and architects. As there was no money in ancient Egypt, they were given food and drink as payment.

Although the pyramids look more or less the same from the outside, not all were built the same way. In some the room for the dead king (called a burial chamber) is below ground, others have the burial chamber above.

The first stage in building a stone pyramid was to flatten the ground and measure out the four sides. Huge stone blocks were cut in local quarries using very simple tools – copper chisels and hammers – and were dragged to the building site on wooden sledges. Ramps allowed the workers to raise the stone to the higher levels of the pyramid.

Once the basic triangular shape had been built it was covered in a layer of very expensive white stone, which was polished so that it sparkled in the sunlight. The top stone of the pyramid – called the 'pyramidion' – was sometimes covered in gold to make it even more shiny!

WHY IS THE SKY DARK AT NIGHT?

Christopher Potter

writer

When we are very young we ask questions all the time. Then, as we grow up, we get embarrassed and stop asking questions so often. Perhaps because we don't want to admit what we don't know. This is sad, because asking questions is really important. Great scientists like Einstein were great partly because they asked questions about things that everyone else thought were obvious.

'Why is the sky dark at night?' looks like a pretty straightforward question. And there is an obvious first answer: because the Sun sets in the evening. But this won't quite do because it suggests that the Sun is moving. In reality, the Sun only appears to move over the horizon. The real motion is of the Earth turning on its axis. This makes it look as though the Sun arcs across the sky. So even this straightforward answer has us thinking about the motion of the Earth in relation to the Sun. And we might easily be led to ask other questions like 'How do we know that the Earth is moving?'

Sometimes a good way to think about a question is to question the question: Is the sky actually dark at night?

If you live deep in the countryside, away from street lights, you will probably have noticed that even on those nights when there is no moon, the sky can be quite bright just from the light of distant stars. For centuries some deep thinkers wondered why the night sky isn't even brighter than it is.

If the universe goes on forever, as many philosophers and scientists think, and if in the infinite universe there is an infinite number of stars, then surely the light from an infinite number of stars should make the night sky really bright. Not dark at all!

But imagine that the universe – space itself – is expanding. In an expanding universe the light from distant stars gets continually shifted further away from us, and that could be enough to explain why the night sky is dark as we see it.

So your question, 'Why is the sky dark at night?' is actually a really deep question to do with whether or not the universe is infinite. And is a question that scientists are still puzzling over.

WHAT SHOULD YOU DO WHEN YOU CAN'T THINK WHAT TO DRAW OR PAINT?

Tracey Emin

artist

I often find I can't make art. At times like this I go and do something else. I usually go partying, play dominoes, go out to eat, or swim, take long walks, go shopping – all the normal things.

I wake up most nights between 1 a.m. and 3 a.m. and stay awake for around two to three hours. That's when I would most like to work but I can't because even though I am awake I am not awake enough to get dressed and go to my studio. But now I have an app on my iPad that lets me draw. The drawings are very different from my usual style because I do them with my finger and I am still a little sleepy so the drawings come from another part of my brain. Also they are very throwaway, so I feel freer.

Reading and swimming are the best things for me to do because the swimming physically makes me happier, and my brain starts working. And the reading fills me with other people's images in my mind, which releases me from stress.

I need to make and create art – I'm an artist. Without creating my life makes no sense, I lose confidence and sort of forget who I am.

HOW DO YOU MAKE ELECTRICITY?

Professor Jim Al-Khalili

scientist and broadcaster

To explain how we can make electricity we first have to know what it is made of. It seems like magic when you really think about it – and even lots of grown-ups don't really know what it is. Maybe you have asked them and didn't get a very good answer. Well, I will do my best here.

The reason electricity is so mysterious is because we can't see it. It just seems to be this invisible energy that makes lights come on and computers, TV sets and pretty much everything else in our world work. I suppose it's a bit like the petrol that a car needs to run. But at least you can see petrol, and smell it, even if you don't know exactly *how* a car engine uses it.

The thing about electricity is that it really is invisible. Not because it's magical, but rather because the things that it is made of are so tiny we can never see them. They are called electrons and they are extremely little specks of stuff that buzz around inside atoms, and atoms are everywhere. Everything in the whole universe, including you, is made up of gazillions of atoms.

Well, electrons carry something called charge, which in a way makes them behave like tiny magnets. The reason electrons get trapped inside atoms is because in the middle of every atom is a powerful atomic nucleus that pulls electrons towards it.

Normally, each atom is very busy coping with this tug of war between the nucleus in its centre and the electrons spinning around the nucleus. So busy that it mostly ignores other neighbouring atoms. The fun starts when some electrons do manage to escape their atoms. They can then march together like an army of soldiers through some materials, like metals, making what is called an electric current. They do this at very high speed.

The reason electrons move like this is because they get pulled by some atoms that are missing electrons and so want to fill their gaps, and, at the same time, they are being pushed away by atoms that are too full of electrons and don't want any more. So billions and billions of these tiny electrons whiz along like this in a wire, and that is what we call an electric current.

OK, so now I've told you what electricity is, here's how we make it.

What we need is some way of pulling lots and lots of electrons off atoms and storing them somewhere, like inside a battery. Then they're ready to be released when we need to make something work, such as a light bulb.

There are lots of ways of generating electricity on a very big scale, but usually it involves getting a special kind of motor, called a dynamo, to spin around, using steam.

Of course, it's not that easy. We need energy to heat up the water that turns into steam in the first place. That energy can come from atoms themselves (called nuclear energy), or from the Sun, or from wind, or just by burning stuff like coal. So there are lots of different steps involved. But at the end all you have to do is flick a switch or push a button and let those little electrons do their stuff.

DID ALEXANDER THE GREAT LIKE FROGS?

Bettany Hughes

historian

Your question has got me scratching my head and thinking all sorts of bizarre thoughts. Well, the Ancient Greek philosopher Socrates famously said 'the unexamined life is not worth living'. In other words, keep your brain switched on and always ask questions of the world – don't just accept things as they are. So thank you for encouraging me to puzzle this over.

Alexander, who was also from Greece (well, actually a place called Macedonia), is famous for many things: trying to conquer the world, fighting battles with elephants and his love of stories by the author Homer. But not many people would automatically put 'Alexander the Great' and 'frogs' into the same sentence. And yet . . . Alexander was taught by the philosopher Aristotle from around 342 BC onwards. Aristotle was one of those Greek philosophers fascinated by *why* stuff happens. He must always have been asking himself questions, a bit like you. Questions like 'Why does a man become a tyrant?', 'How does a block of stone become a sculpture?' and 'Why do tadpoles end up as frogs?'

Aristotle himself was taught by Plato, another great thinker. Plato once said of the Greeks, 'We live like frogs around a pond', because for the Greeks much of life – fighting, shopping, exchanging ideas – involved travelling across the Mediterranean sea. Aristophanes, a playwright from Athens, had a great success with his comedy called *Frogs* (written in 405 BC). And one of Aesop's fables, called 'The Boys and the Frogs', is about some mean boys throwing stones at frogs in a pond, meaning that what we do in fun often causes trouble for others.

So obviously these men in Ancient Greece actually spent a great deal of time thinking about frogs and talking about them. Why should Alexander be any different? Alexander loved Homer (he slept with a dagger and a copy of Homer's book the *Iliad* under his pillow). So there's the strong possibility he might also have been aware of the comic epic *Batrachomyomachia*, 'The Battle of Frogs and Mice', which some people thought was by Homer.

And there's no doubt Alexander's experience of frogs would have extended beyond just reading about them. If you spend time in the Mediterranean away from the sounds of the twenty-first century (cars, trains, planes, mobile phones) and walk through the countryside, frogs make their presence felt in no uncertain terms, croaking and singing away in chorus. It can be like a frog opera out there. Thank you for your question – I'll never think of Alexander in quite the same way again.

WHAT ARE OUR BONES MADE OF?

Professor Alice Roberts

anatomy expert and broadcaster

Bone is amazing stuff. You might think of bones as being white, brittle and lifeless, but the bones inside your body are very much alive.

Bones are made out of a very hard material but there are lots of tiny cells inside that material. Bones are also pink, because there are so many blood vessels in them. And bones are incredibly strong – as tough as iron, but they're not brittle. It's actually quite hard to break a bone, luckily. This is because the bone material is a very clever mixture of hard mineral, which contains a lot of calcium, and tough protein.

Bones are always changing, on the inside and on the outside. When you're still growing, they're obviously changing shape and size, but even when you're an adult, they can still change a bit. This is because they contain living cells. Some of these cells, called osteoblasts, can pump out new bone material. Others, called osteoclasts, eat away at bone material. Together, the osteoblasts and osteoclasts make sure the whole

189

bone is always the shape and size it needs to be, to stand up to the forces you put it under.

If you took a bone like your thigh bone, or femur, and sliced it in half, you wouldn't see the cells. (You need a microscope for that.) But you would see a difference between the type of bone in the middle and at the ends. In the middle of a bone like your thigh bone, the bone material is arranged like a thick cylinder or tube, and the inside of the bone contains marrow, which is mostly fat in an adult, but has blood-making cells in it while you're young.

The ends of the bone look different: there's no marrow cavity in the middle, they're completely full of a type of bone that looks like sponge. In fact, it's called spongy bone. Of course, it's not soft and spongy, it's actually very hard.

Because bone is alive, and full of cells and blood vessels, it's very good at mending itself when it gets broken. It helps if you can keep the broken ends of the bone still, which is why doctors will put a splint or a plaster cast on a broken arm or leg. After just a few weeks, new bone will have grown to 'glue' the broken bits back together. I hope you'll agree – bone is amazing!

IF YOU'RE ON A BOAT WITH NO FOOD OR WATER, WHAT DO YOU DO?

Roz Savage

the first woman to row three oceans

Luckily this has never happened to me. I always take heaps of food with me, and I have a machine that makes drinking water out of sea water. But if I ran out of food and my water-maker broke, I'd have to get creative.

For food, I could catch fish. But I'd really hate having to do this. I get a group of fish gathering under my boat, and as time goes by I get to recognise individual ones by their size or by scars on their skin. I'm all alone on my boat, so the fish are the nearest thing I have to company. Sometimes I even talk to them. (If they ever start answering me back, though, I'm in trouble.) I would find it really hard to catch and kill them, but if I was hungry enough, I suppose I'd have to do it.

As for water, I would have to catch rain using my sun canopy. But this would be quite difficult. Often I don't get any rain for days or even weeks. Or if it does rain, sometimes it is so windy that the rain is flying horizontally and it would be hard to catch it. So I might also have to try and spot a passing ship and ask them for water. I just hope they wouldn't give it to me in plastic bottles, because I see a lot of plastic rubbish floating in the ocean so I try to avoid using bottled water.

Generally, I'll just carry on being well prepared for my voyages, and hope that I never end up with no food and no water. Life on the ocean is hard enough – with waves that are sometimes big enough to capsize my boat, and storms, and sharks. So to be hungry and thirsty as well would be really just too much!

HOW DOES MY CAT ALWAYS FIND HER WAY HOME?

Dr Rupert Sheldrake
biologist and author

If she finds her way home over quite short distances, from places where she's been before, she is probably just remembering familiar landmarks, buildings, trees and so on. Just like you would if you were going home from a familiar place. But some cats find their way home from many miles away over unfamiliar territory, after people on holiday have lost them and have had to go home without them.

Dogs do this too. They seem to have a sense of direction that helps them to find their way back from places they have never been before – in some cases from hundreds of miles away, as shown in a Disney film called *The Incredible Journey*, which was based on a real-life story.

This is just the tip of the iceberg of direction-finding abilities in animals. Homing pigeons find their lofts from far away; they do this for pigeon races all the time. Racing pigeons can fly home from six hundred miles away in a single day. They can't do it by seeing their home, and scientific research has shown that they don't do it by remembering the twists and turns of the outward journey. Nor does it all depend on the sun's position, because they can fly home on cloudy days too, and can even be trained to do it at night.

The Earth's magnetic field seems to play some part in their homing ability. A compass points north because of this magnetic field, so you can use it to tell which direction you are going in. But even if the pigeon had a compass sense, that could not fully explain it. If you were parachuted into an unknown place with a compass, it would tell you where north is, but not where home is.

Migrating animals and birds achieve even greater feats of navigation. British cuckoos migrate to southern Africa, crossing the Sahara desert and leaving their children behind. The young cuckoos left in Britain are raised by birds of other species, and never meet their parents. Yet several weeks after the older generation has left, the young cuckoos join together and find their way to their parents' home region back in Africa.

Again, magnetism seems to play a part in migratory animal behaviour but it is not the whole explanation. I myself think that animals are connected to their homes by a field that acts like a kind of invisible elastic band. When a pigeon is released hundreds of miles from home, it circles around and then heads off homewards, as if responding to a pull. Young cuckoos inherit their sense of direction, and seem to be pulled by an ancestral field, a kind of collective memory in the species. But this is just a theory. No one really knows how animals do it.

WHAT'S INSIDE THE WORLD?

Professor Iain Stewart

geologist

Rock. Over six thousand kilometres of the stuff! That's roughly the distance from Paris in France to Delhi in India, except straight down to the centre of the Earth.

Way down in the inner core of the Earth, the humungous pressures of the planet above have squeezed metal-rich rock into solid iron. If you could go down there, you'd find individual iron crystals many hundreds of metres in length.

Further out, where the pressures are less but the temperatures are still hotter than the surface of the Sun, that same material flows as liquid iron. It is the swirling of this turbulent ocean of iron – the outer core of the Earth – that generates our planet's magnetic field and keeps the surface parts of the planet in motion.

Imagine a vast boiled egg in which the yolk is only partly hardened – the runny yellow bit is like this fluid outer core. If you keep that image in your head, then the rubbery white of the boiled egg resembles the lighter rocky materials that form the surrounding bulk of the planet. This is the Earth's 'mantle'. Here, at depths of many hundreds of metres, the mantle rock is easily hot enough to melt, but intense pressures keep it solid, or at least the kind of solid that squishes like warm plasticine.

Above this is the wafer-thin shell of the planet: a tough, brittle 'crust' that is usually a few tens of kilometres thick.

Only in the Earth's crust do temperatures finally drop to less than a hundred degrees centigrade. And the constant loss of heat from the super-hot interior means that Earth's cold, rigid shell is cracked from below into a shifting jigsaw of broken fragments. We call these 'plates'.

In places where the plates break apart, the release of pressure makes mantle material (the 'egg white') immediately underneath melt suddenly and escape upwards. It bursts out as molten lava from volcanoes.

Volcanoes burst up most easily through the ocean floor, where the Earth's crust is thinnest. As these fiery cracks cool, new crust is born. Elsewhere, crust is destroyed, lost in those places where plates collide and crumple or one slides beneath another. The scars left behind from this huge recycling scheme are the great mountain ranges like the Himalayas and the Andes. In fact, wherever you look on the face of the Earth – whether it is continents and oceans or mountains and volcanoes – you see the result of millions of years of moving plates.

But what is really amazing is that the engine that powers this spectacular planetary machine lies thousands of kilometres below, in Earth's partly molten heart.

196

WHO IS GOD?

This was one of the most frequently asked questions – and it's clearly a question with many possible answers. We asked three Big People, each with a different view on the subject. Their answers follow.

Julian Baggini

philosopher

Who is God? It's a good question and the truth is that everybody seems to have an idea of who he is but nobody really knows.

For many, God is a bit like a father, but a father to everyone. He created the universe and everyone in it, and he loves all of us. But he is also prepared to tell us off and punish us if we do the wrong thing. People who believe in this God think we should obey him and love him like we obey and love our human parents.

But these people disagree among themselves about who exactly God is. Because people have different ideas there are many different religions – and different groups within different religions.

Other people think God isn't a person at all but a kind of force. The world is full of good and evil and 'God' is the name we give to the good.

Yet others think that God doesn't exist. Human beings invented the idea of God to explain how the universe started and why we should be good. But now that we have science to help us understand the world better, these people think we don't need to believe in God any more.

So there is no simple answer to the question 'Who is God?' You will have to work out which answer makes most sense to you. As you do, my personal advice would be this: if anyone tells you they know for sure who God is, be suspicious.

Meg Rosoff

author

Now there's a question! Is God a man? Or a woman? Or a fish? Or a goat? Is God old or young? Fat or thin? The size of an onion or a dinosaur or Mount Everest? Is God as slow as a snail or as fast as a shooting star? Is God invisible? Out to lunch? Listening carefully? Or just an idea someone thought up ten thousand years ago?

Does God live in heaven? On a cloud? Somewhere in outer space? In our heads? In the Bible? Or no place at all?

Some people think that God created people.

Some people think that people created God.

Some people think that their god is the only god.

Some people think there are lots of gods – hundreds of them!

Some people would kill anyone who disagreed with them about exactly who or what God is.

Some people are absolutely positively one hundred per cent certain that there is no God.

Some people just . . . aren't . . . too . . . sure.

Maybe God is a feeling. A nice feeling that makes you feel safe. Or a shouting feeling that says Thou Shalt Not do this, Thou Shalt Not do that, Thou Shalt Not have any fun at all! Maybe God is the voice in your head that tells you not to hurt other people. Not to steal or kill or lie to yourself. Or to put the top back on the mustard.

Maybe God is like nature. Like a sunny day or a wave in the ocean. Maybe you only see God when you need to see God. Or maybe God isn't there after all.

No one can tell you that your God isn't the right God, or that your idea of God is the wrong idea.

You don't have to believe in God. God doesn't have to believe in you. It's your decision. And you can always change your mind.

Francis Spufford

author

First of all, here's who God isn't: He isn't a superhero. He isn't somebody like us, only stronger and faster and cleverer, using

His special powers to zip around the world. In fact, He isn't part of the world at all. If you believe in Him, He's the reason there is a world. All the things you see are here because he pours in love to keep them going.

You can't prove He exists. (And you can't prove He doesn't, either.) But people who believe in Him – like Christians, Jews and Muslims – tend to think we can feel Him being there. For us, He's there in the peaceful stillness of our minds, He's there in the sound of prayers, He's there when we don't feel lonely on a lonely road. Christians tend to feel He comes closest when we're being loving, and Jews and Muslims tend to feel it's when we're behaving fairly, but we all agree He cares about us, and He cares what we do.

We make mistakes and get things wrong, but He never gives up on us. He's the person who loves us no matter what. If that sounds like an ideal mum or dad, then it's not surprising, because to people who believe in Him, he is the mum and dad of the whole universe. Maybe we invented Him by thinking of mums and dads and then imagining a very big one, but it doesn't feel like that. It feels more as if the good things about families are a kind of little glimpse of what the universe we live in is really like in the end, despite everything.

When we do cruel things or destructive ones, we get further away from Him, and when we do kind things or sympathetic ones, we get closer to what He is like. Compared to Him, we're very temporary little people, looking at the world through the tiny windows of our two eyes. But oddly enough, thinking of

Him doesn't make us feel small – at least not in a gloomy or discouraging way. It's more like what you feel if you climb to the top of a very tall mountain, where the sun glitters like a diamond in a dark blue sky, and you can see for hundreds of miles in all directions. You discover that the world is much bigger than you knew it was, and that maybe, just maybe, you can be bigger than you thought you were, too.

HOW MANY DIFFERENT TYPES OF BEETLE ARE THERE IN THE WHOLE WORLD?

Dr George McGavin

entomologist

Today there are about 387,000 named beetle species. People only started naming and classifying species properly about three hundred years ago and since then around 1.5 million species of animal have been described and named. Of these, around one million species are insects and the most numerous kind of insect is a beetle. Put another way, there are more sorts of beetle on Earth than anything else.

But we cannot be sure of the exact number. Sometimes a single species may be named more than once by mistake and more are being discovered all the time. You might also ask why there are so many insects and particularly why there are so many beetles. Well, insects have been around for more than four hundred million years and are very successful because they are generally small and breed very fast.

Insects were also the first animals to take to the air. They were flying millions of years before birds or bats appeared.

Like many insects, beetles have two pairs of wings but the front wings are toughened. These wing cases or 'elytra' protect the larger and more delicate hind wings when they are not in use. This allowed beetles to colonise all sorts of different places on the planet.

Then, about one hundred million years ago, the evolution of flowering plants gave beetles a whole new range of places to live and things to eat, and the number of beetle species increased dramatically. While there are many more beetle species to discover, especially in tropical forests, we may never know about them as these habitats and the animals that live there are being destroyed.

HOW FAR AWAY IS SPACE?

Marcus Chown

author of books about space and the universe

Probably, you think space is thousands or even millions of miles away. Actually, space is only about twenty miles from your doorstep – straight up. Almost certainly you could walk twenty miles, though you would get very tired and likely moan a lot. But to go twenty miles straight up, you need a rocket.

Rockets are actually very bad at getting into space. The trouble is there is no rocket fuel strong enough to boost itself plus the metal skin of a rocket into space in one go. The only way people can get a rocket into space is by throwing part of it away when the rocket is high up in the air. This makes what is left of the rocket lighter and so it's easier for the fuel to boost it all the way into space.

Imagine if every time your mum or dad drove to the super-market they threw away most of the car, leaving only a steering wheel and four tyres. So that the next time they went to the supermarket, they had to rebuild the car. Ridiculous? But that is just what they have to do for rockets: rebuild them for each launch. No wonder it's so expensive to get into space – about $500 million for every launch of the American Space Shuttle.

The sensible way to get into space would be to build a ladder twenty miles high. Unfortunately, a ladder that high, even made out of the strongest metal we have, would crumple under its own weight. But stronger materials are being invented all the time. There is a good chance that you will see something like a space ladder – more commonly known as a 'space elevator' – built in your lifetime. Then, at last, it will be cheap and easy to get into space. We may even go there for our holidays.

HOW DOES LIGHTNING HAPPEN?

Professor Kathy Sykes

physicist

Watching lightning in the sky can be spectacular. It looks so mysterious and even now we don't know everything about it.

We know that it usually happens in 'cumulonimbus' clouds, ones that are incredibly high, even sometimes reaching over twelve miles. These clouds form during thunderstorms. They are often dark and angry-looking and sometimes have an 'anvil' at the top – an area of cloud that extends a long way from the middle of the cloud and can look like the top of a mushroom.

Within cumulonimbus clouds there are very strong winds. (So strong that it's dangerous for small planes to fly into these clouds.) The winds carry moist air upwards into the cold areas high up, and rain and ice particles form. We think the rain and ice and winds inside the cloud may be what makes lightning form. But before we talk any more about how lightning actually happens, we need to understand a little about atoms.

Everything is made up of atoms. You are made up of atoms, so are rocks, water, animals, plants and air molecules. Atoms contain positive charges, balanced by negative charges, which are called electrons. Usually, these positive and negative charges stay closely bound together because they attract each other strongly. But large forces can separate them. And once the two charges are separated, they 'want' to get back together again as soon as they can.

Now let's go back inside our cumulonimbus cloud. The rain colliding with the ice particles, driven by the strong winds, may be what separates negative charges from positive charges. Negative electrons gather at the bottom of the cloud, and positive charges are swept up to the top of the cloud by winds. Exactly how the charges separate is not totally understood. Scientists have different theories. But once there is a build-up of negative charge at the bottom of the cloud and positive charge at the top, there is the possibility of lightning forming. The charges want to come back together again. The strong negative charge at the base of the cloud wants to become neutral again by connecting either with the positive charge at the top of the cloud or with the ground below it, which is relatively positive.

Eventually, the charge differences are so great that the electrons actually start trying to get to the ground. A 'stepped leader' forms – this is the name of the first streak of lightning coming from the cloud, usually over fifty metres long. This will branch, and more stepped leaders form. As these get closer to the ground, positive charges on the ground want to connect to the strong negative charge of the tip of the lightning.

If you ever feel your hair beginning to stand on end in a lightning storm – start worrying! It will be because the positive charges on you are wanting to connect with the negative charges in the cloud, and you are being attracted to a stepped leader. Your hairs are able to move and show when charges on you are trying to move towards something.

After a short time, the stepped leader reaches the ground, or positive charges from the ground will reach it. Then the lightning has struck, and pulses of electric charge go to and from the cloud. It's the positive charges going from the ground, called the 'return stroke', that are actually the really bright parts of the lightning. The stepped leaders are almost invisible.

Sometimes you can see lightning within clouds or between them. The negative charges from the bottom can form stepped leaders, which go from the bottoms of clouds to the tops.

Either way, this amazing scene is the atmosphere's way of trying to balance the charges that have been separated.

WHY ARE SOME PEOPLE TALLER THAN OTHERS?

Katie Woodard

forensic scientist

Everyone has 'DNA' in their cells (what all living things are made of). Your DNA is from your mum and dad, and it is like a magical code that stands for everything that happens in your body, from the day you started growing in your mummy.

As you have probably noticed, some races (big groups of people) can be taller or shorter than others overall. This is because they have evolved over time to be this way – we're talking over thousands of years! This is based on many reasons, such as how much healthy food they were able to get that whole time.

But there's more to the story! Just because you plant a flower seed that doesn't mean it will grow into a perfect beautiful flower, does it? Just like a flower needs sunshine, water and good soil to grow, you need certain things to be as tall as you are meant to be – as tall as your DNA codes allow. For people this means getting enough sleep, exercise, and most importantly eating healthy food – fresh, whole foods, especially home-made, with the most nutrients in them.

WHY IS WEE YELLOW?

Sally Magnusson

journalist

Wee starts out as blood that has finished its work. It ends up by helping us to do some amazing things.

Imagine blood as a train chugging around our bodies, picking up and dropping off all sorts of vital cargo to keep us healthy as it goes. At the end of each journey there are always a few bits and pieces left over. In the blood's case that includes thousands of important chemicals, such as nitrogen and ammonia.

Our kidneys send these leftovers straight to the bladder, along with lots of spare water. Out they all whoosh a few times a day – and that's wee.

But why is it yellow? Well, the cells that give blood its red colour eventually get worn out by all that chugging around. As they die off, they turn yellow. They are then turfed out in our wee as well, turning it yellow too. That yellowy golden colouring is called urochrome.

But you may have noticed that wee is not always yellow. Some foods leave their own colour. Check your wee after you have eaten a lot of beetroot. It's bright red. Eat too many carrots and it can look a bit orange. Asparagus may give it a greeny tinge.

And if you don't drink enough, your wee will become dark. This is serious. In fact, from earliest times doctors have examined the colour of people's wee to work out what is wrong with them. Poor King George III, who had a mental illness, managed to produce blue wee – which must have been quite a shock.

And here is the amazing thing. Remember all those important chemicals in our wee? They can be used again.

All through history people have been rubbing wee into their skin to heal wounds and soothe burns. Others soaked plants in it to make dye. It has been used to make bread (yes, really). It helps flowers and crops to grow. For centuries, believe it or not, it was a vital ingredient in gunpowder.

The ammonia in wee means it can also clean just about anything. The Romans washed their togas in it and until recently weavers used it to clean cloth. People in Britain used to be able to sell theirs for a penny a bucket. Don't get your hopes up today, though!

Mind you, wee is still proving useful. Scientists in Scotland have discovered how to generate electricity from it. In Denmark pig wee is being recycled to make plastics and – wait for it – lipstick. In the USA researchers are producing hydrogen from wee and hope that one day it might even power cars.

Oh, and it makes great invisible ink.

Not bad for a humble yellow liquid we keep flushing away, is it?

WHAT WAS THE BIGGEST BATTLE THE ROMANS FOUGHT IN?

Gary Smailes

military historian and author of books for children

Let's face it, Romans were just big SHOW-OFFS with their shiny armour, pointy spears (called a *pilum*) and lethal, razor-sharp arm-and-leg-cutting-off swords (called a *gladius*). The Romans were also a pretty ungrateful bunch and were only happy when beating up other nations. In fact, Romans were a bit like school bullies, but with metal helmets, scary swords and strappy sandals.

This meant that the Romans had LOADS of battles. The question is, what was the *biggest* battle?

Now, here we have a problem. Not only were the Romans bullies, they were also MASSIVE fibbers. After a battle their historians would want to make the Romans look as hard as possible. So they would tell people that there were LOADS of the enemy and almost NO Romans. The Roman historians would be all, 'Oh yeah, there were just four Romans and two of them were a bit tired, and there were, well . . . SIXTEEN BILLION of the enemy!'

Luckily, today's historians have become pretty good at spotting these lies and working out what really happened.

This means that, though we don't know for sure, we think that the BIGGEST Roman battle was the battle of Philippi. That battle took place forty-two years before Jesus was born. It all started when the Roman ruler, Julius Caesar, became involved in a little stabbing incident and was brutally murdered by his so-called mates. After Caesar's murder everyone was a bit peeved (well, not as upset as Caesar, but you get the idea) and they decided to have a HUGE fight to decide who should be the new Roman ruler.

On one side were Mark Antony and Octavian. They decided they needed a cool name and called themselves the Triumvirs. On the other side were Marcus Junius Brutus and Gaius Cassius Longinus. They felt a little left out and also decided to give themselves a cool name. They called themselves the Liberatores.

Soon after the whole stabbing-Caesar-to-death thing, Brutus and Cassius legged it with their armies to Greece, where they hung out for a bit in the sun. Mark Antony and Octavian also had armies and decided to march to Greece. They all met up just outside the city of Philippi.

Today we believe that the Triumvirs had about a hundred thousand soldiers, but if you count all the other possible troops and people helping out it was probably about 223,000 men. The Liberatores also had about a hundred thousand soldiers, but this goes up to 187,000 if everyone is included in the numbers. That means that there could have been about four hundred thousand men involved in the battle. That's enough to fill Wembley Stadium four times over, with a few people still left outside buying hot dogs.

So what happened in the battle?

The battle was actually TWO battles at the same time. In one Brutus (Liberatores) faced Octavian (Triumvirs). Brutus was the better general and pushed back Octavian's army. This meant it was 1–0 to the Liberatores.

In the other battle Cassius (Liberatores) fought it out with Mark Antony (Triumvirs). Here Antony won, making it a 1–1 draw. However, someone told Cassius the naughty lie that his mate Brutus had lost (he had actually won). Cassius got in a BIG mood and killed himself.

With the battle a draw they had to have a replay. This time Brutus and his army were all alone and he lost the battle. Afterwards, Brutus was a bit of a copycat and also killed himself. The prize for the Triumvirate was the Roman Empire.

WHY DO I GET BORED?

Professor Peter Toohey

academic and author

You know what elephants are like. They are big and grey and very strong. And they have very, very long grey, hairy noses called trunks. They can pick things up with their long noses and they can suck things up with them as well.

I don't think I would ever get bored if I had my own trunk. I'd use mine to suck up water and spray my friends for fun. But elephants do get bored. And when they get bored it makes them grumpy. They sway from side to side, thumping around on their great big legs, and they toss their trunks all over the place.

How do you cure elephants of boredom? You play them some music. They like serious, old-fashioned music with lots of violins. That doesn't surprise me, because I've always thought elephants were very old-fashioned things. They live for a long time and become very old. Do you like the sort of music that elephants do? I'll bet you don't. You're probably more like a chimpanzee. Some scientists at Belfast Zoo in

Northern Ireland have discovered that chimps get over being bored and grumpy if they listen to rock 'n' roll.

But why do elephants get so bored that they need to listen to music? They get bored if they're in small zoos and there isn't enough to do. They get bored if they can't wander about with their friends, and if they know exactly what's coming up next: hay for breakfast, hay for lunch, hay for dinner. Same bed, same old cage, same old friends.

You get bored the same way. There's not enough to do. Your friends are somewhere else. You have to be still and quiet and stay indoors when you really want to play outside.

Being bored is your body telling you to do something different, so you don't get sad or grumpy. You need to get out with your friends and family, and find new and exciting things to do. Next time you feel bored, why not try the elephants' cure? Put on some music and swing your trunk. Or be a monkey, and listen to some rock 'n' roll!

ARE THERE REALLY MONSTERS LIVING IN OUR MOUTHS CALLED BLACKTERIA?

Liz Bonnin

science and nature TV presenter

There are no monsters living in our mouths but what does live there is far more interesting. Our mouths provide the perfect environment for hundreds of different types of micro-organisms like bacteria, viruses and fungi to exist.

So many, in fact, that microbiologists haven't even identified all the different types yet. They are far too small for the eye to see and they live quite happily in different parts of our mouths – on the cracks in our tongues, in the spaces between our gums and teeth, and on the roofs of our mouths. As many as a hundred thousand of these fascinating creatures can be found living on just one tooth.

The bacteria found in our mouths live in communities called biofilms, and they can communicate with each other and with other species of bacteria as they colonise a tooth or invade a new area.

Some of these organisms may look a little scary when they

are magnified thousands of times by a special machine called an electron microscope, but many of these living things can actually do us good. They protect us from bad types of bacteria, eat up the food lodged in our teeth and make different products that can actually help keep our mouths healthy.

Our body's natural defence system is excellent at keeping the numbers of these tiny organisms in check, so that they don't get to a level that might do us harm. And if you make sure you brush your teeth and keep them minty fresh, you're also helping to ensure that all of them remain nicely balanced and don't cause any mischief.

But we've all heard of cavities and we get them because some types of bacteria can do damage if we don't take care of our teeth and gums. Two of the best-known have great names: *Streptococcus mutans* and *Lactobacillus acidophilus*. They make acids when they feed on the sugars that we love to eat, like those found in sweets and chocolates.

Now, normally our saliva gets rid of the acid produced by these bacteria and there isn't a problem. But nowadays we eat so much of these refined sugars that it's like Christmas every day for the bacteria. With all this sugar in our mouths, the bacteria produce so much acid that our saliva can't deal with it all. This erodes our teeth and causes nasty cavities. This is why we need to visit the dentist much more than we did in the past, when we hadn't figured out how to refine sugar and put it in so much of our food.

But as long as we brush and floss our teeth regularly, we can prevent these cavities and other problems from developing.

And we can keep the rest of the thriving micro-organisms at a nice healthy level.

Just as we have been able to populate the planet, we have returned the favour by giving all these incredible micro-organisms a great place to live in our mouths. Which is quite nice when you think about it!

WHY DO WE SLEEP AT NIGHT?

Russell G. Foster

professor of circadian neuroscience

We sleep at night because our bodies are adapted to be active during the day. Other animals, such as bats or badgers, sleep during the day and are active at night because this is when they hunt to find their food.

We have good eyesight when there is plenty of light, but at night we see poorly and find it difficult to get around. Bats and badgers have poor eyesight, and use sound and smell to find their way around at night. But this does not explain how our pattern of sleep is controlled.

The brain tells us when to sleep. Deep in the brain is a biological clock consisting of about fifty thousand nerve cells that work together and act a bit like an alarm clock, telling the rest of the body what to do at different times of the day, and when we should be asleep or awake. Tiredness is also controlled by another part of the brain, which measures how long we have been awake. The longer we are awake the more tired we feel.

Flying to other countries thousands of miles away in different time zones gives us jet lag. When it is daytime in Australia it is night-time in England, and when we are going to bed in England, people are getting up in California. Our body clock cannot adjust to the new time zone instantly. It takes several days. So you feel tired or hungry at the wrong times in Australia or California until the clock in the brain moves from home time to the new time zone. We recover from jet lag because light in the new time zone, detected by the eye, regulates our body clock.

So the body clock and levels of tiredness work together to regulate our sleep patterns. Many people think that the brain is turned off during sleep, but this is wrong. Some parts of the brain are even more active during sleep than when we are awake! This is because during sleep the brain is helping us to remember what has happened during the day and to make sense of new information. Many people wake up in the morning and find they have the answer to a problem that has been puzzling them for ages!

The rest of the body also undergoes lots of changes while we are asleep. Young people grow more during sleep than when they are awake, and damage to the body is often repaired at night. When we are young we need about nine hours of sleep each night for the brain to be fully active during the day.

You are better at solving problems, less moody, better at sport and will even find jokes funnier with a good long sleep. Many grown-ups do not get enough sleep, sleeping only five or six hours each night. If this goes on too long, they can become seriously ill with conditions that can affect their digestion or heart and they might even suffer from depression.

For a long time we did not realise why sleep is so important. Now we know that lots of helpful things are going on in our bodies during sleep. Sleep helps make us both healthy and happy. So make sure you get enough sleep!

WILL WE EVER BE ABLE TO GO BACK IN TIME?

Dr John Gribbin

author of science and science fiction books

Time travel is possible, but it would be very difficult to build a time machine. You might need two black holes to do the job! The rules of physics that describe how space and time work tell us this – rules that were worked out by Albert Einstein in his General Theory of Relativity.

A black hole is like a hole in space and time, and if you had two joined by a time tunnel you might be able to jump in one and come out of the other one at a different time. Saying time travel is possible is a bit like telling a Stone Age person that space travel is possible. They wouldn't be able to do it until they had learned how to build the machinery first.

There's another snag. The rules also tell us that it would be impossible to go back in time to some time before you built the time machine. This is like the way you can't go anywhere on the London Underground that doesn't have an underground railway line. It makes sense, because there wouldn't be a time machine in the past for you to go back to! The 'other end' of the black hole is stuck in the day it was made.

So if somebody built a time machine tomorrow, you could use it to go to any time in the future and come back to tomorrow. But you could never go back to yesterday. And that explains why the world is not full of time-travelling tourists from the future – proof that nobody has built a time machine. At least, not yet. Your only hope of going back in time from today is if you can find a time machine that somebody has built already.

If you did find one, where would you like to go? I'd like to go back a hundred years to meet Einstein, the man who explained how space and time work.

HOW DOES FIRE GET ON FIRE?

Dr Bunhead

stunt scientist

I am NOT going to tell you how to make a fire because it is a HUGE SECRET. Also burning things can be REALLY DANGEROUS! You might set fire to your friend or your socks or something else super-bad.

But if I did tell you then you would have to keep it to yourself. You cannot tell anyone else, except maybe your best friend or your pet tarantula. You have to promise to write the recipe down very carefully. Then fold it up really tiny. Then wipe bogeys all over it. Then put it in a jar of old toenails so NO ONE else will dare to read it. Promise?

OK then, I am prepared to share the Scientists' Secret.

Recipe for a fire
Make sure no one is looking. Here goes . . .
1) Some FUEL
2) Some HEAT
3) Some AIR

Yes that's it. You just need THREE ingredients to make fire. But you also have to know more science stuff. So you will have to read the next bit and it's VERY BORING (unless you like fires, then it's very interesting).

Secret Ingredient Number 1: FUEL
FUEL is the stuff that burns. Wood, paper, oil and coal are all good fuels. Things like hands, rocks, paperclips or bogeys are not good fuels.

Secret Ingredient Number 2: HEAT
HEAT is what you need to start the fire. You get heat from things like a spark or rubbing things together really fast or even shining sunlight through a magnifying glass. You can probably think of lots of other hot things you could use.

Secret Ingredient Number 3: AIR
You must have AIR for a fire to burn. Actually you need something that's in the air. But that's such a huge top secret you would have to read right to the end to find out.

Making your very own fire
Rub your hands together. Faster. As fast as you can! Feel how hot they are! Have they caught fire yet? No? Don't worry, your hands won't catch fire. Even if you made them really, really super-scorching hot. Remember, hands are not good fuel, they are rubbish at catching fire.

232

To make your fire first you need a good FUEL. A small bit of dry wood is a good fuel. Then you have to make your fuel really HOT. You could make wood really hot by rubbing it very fast against another bit of wood. Last of all, add a bit more AIR by blowing gently over it and WOOOOMPH!! There's your fire.

Once something catches fire it gives out its own heat and so it gets hotter and hotter. So hot it can catch other nearby things on fire until everything around it is on fire. That's why we have to be VERY CAREFUL when making fires. If we know how to start a fire we should also know how to PUT FIRES OUT. But that's a question for another time.

Super Science Secrets
1) The proper scientific word for things that burn is FLAM-MABLE ('fla-ma-bull').
2) 'Flammable' and 'inflammable' both mean the same thing. Silly but true!
3) The proper scientific word for burning is COMBUSTION ('com-bus-chun').
4) Air has lots of different gases in it. The gas we need for burning is the HUGE TOP SECRET . . . Shhhhh. It's called oxygen ('ox-i-jun').

WHY DO WE HAVE LOTS OF COUNTRIES, NOT JUST ONE BIG COUNTRY?

Dan Snow

historian

Even though humans are all the same, our ancient ancestors travelled so far across the world, from Tasmania to Timbuktu and from Alaska to Aberdeen, that they developed differently. Over thousands of years they got different skin colours and languages, they invented different religions and lifestyles.

A few thousand years ago humans started to invent countries. They invented places like China, Japan and Egypt. The trouble was that the people in China did not know that the people in Egypt even existed because there were no cars, trains, planes, phones, internet or even big ships. So they did not have the chance to get in touch with each other and agree to just have one joint country.

By the time all the countries realised that the other ones existed, only a few hundred years ago, lots of people did not want to join the countries together. The kings, queens, emperors and chiefs of all the countries did not want to join another country because they did not want to share their power or their palaces with anyone else.

They encouraged their followers to stay separate from other countries. Those followers usually agreed with their ruler. They did not like or trust the people they met from other countries because they spoke strangely, ate weird food, worshipped a different god, and even looked different. They wanted to keep their own country because they were used to it. They understood it. If they joined with another country everything might change. And change makes many humans very scared.

Some people thought having lots of little countries was stupid. They thought it would be much better if everyone lived in one country, preferably one that they were in charge of. They would send an army to attack another country and take it over. But often the people who had been taken over wanted their old country back, because they did not like having new, foreign people in charge and they were angry that they had been attacked and their friends killed or hurt.

Today we can travel right across the world, or speak to someone over the internet no matter where they are. We can eat the same food in Shanghai or Sunderland. Soon computers are going to be able to translate one language into another as fast as we can say the words. We have a lot more in common with other people in other countries than our ancestors did. At the United Nations or in the European Union countries agree to work closely together and make laws and establish rights that are the same across lots of countries. Perhaps we are slowly getting closer to living in one big global country.

236

WHAT MAKES ME ME?

This was one of the trickiest questions sent in. We asked an expert on early humans, a psychology professor and a children's author what they thought.

Professor Chris Stringer

palaeoanthropologist

If you watch grown-ups cooking a special meal, they go and find the ingredients, like meat and vegetables and spices, and maybe use a recipe from a cookbook by Jamie or Delia or Nigella.

If you think of your body as like that special meal, the ingredients are all the chemicals and little cells that make up your body and make it work.

The recipe that said how all the ingredients in your body would be prepared, put together and cooked in different ways, is called your genetic code. It's like a tiny but very long book of instructions for how to make you. This genetic code was in the egg that began your life, inside your mum.

Our genetic codes (recipes) are all a little bit different from each other, with slightly varied lists of ingredients, and slightly different instructions for how to prepare those ingredients. Just as there are lots of distinct kinds of curries, because of the endless combinations of ingredients and ways of cooking them, there are lots of different kinds of people because of the slightly different recipes that have made us all.

That is why you are you. Why you are the shape and size and colour you are, and why (unless you have got an identical twin, with a very similar genetic code or recipe) there is no one else like you in all the world!

Professor Gary Marcus

cognitive scientist and author

What makes you you? Just about everything that you can think of: your head, your arms, your toes, your heart and most especially your brain.

If you lost a toe, of course, in some rather unfortunate accident, you'd still be you, just 'you without a toe'. The same goes, I suppose, for your left arm, or your right kneecap, though I'm sure you'd miss them both.

Your brain, however, is a different matter. If there is one part of you that most makes you you, it's probably that: your brain, the three pounds or so of 'grey matter' lodged inside your skull that help you think, reason, and remember.

Without your brain, you wouldn't know how to get out of

bed in the morning. You wouldn't have any ideas. And you wouldn't remember who you were; you wouldn't even be able to ask the question 'What makes me me?'

All of which raises *another* question: what makes *your* brain your brain? You can go to the shop to pick out a new shirt or new pair of shoes, but the brain that you have is the brain that you are born with. Even your heart could be replaced, but if you replaced your brain, you wouldn't be you any more. Your whole personality might change if you did! For it is your brain that makes you happy or sad, nice or mean, friendly or shy.

Your brain started to become what it is when you were still inside your mother's womb. A sheet of cells, sort of a layer of skin, folded over on itself and formed a tube. That tube began to puff out, and eventually divided into two halves (called hemispheres). Then it further divided into sections, like the frontal lobe, which helps you make decisions, and the temporal lobe, which helps you understand the things that you hear.

Much of your brain's basic shape originally came from your parents, by way of their genes. But ever since then, it's been up to you. Because every time you try to learn something new, your brain changes. You can't order a new brain online, but by learning something new every day, you can keep making the brain you already have even better.

Because no two brains are quite alike, no two people think alike, or act alike. More than anything else, it is *your* brain that makes you you.

239

Michael Rosen

author and poet

I look at my parents and say, what have you given me? I look at my grandparents, uncles, aunts, cousins and say, what have you given me? I look at the schools and clubs I've been to and say, what have you given me? I look at the places I've been and stayed in and say, what have you given me? I look at my friends and people I've loved and say, what have you given me? I look at the plays I've been to, the books I've read, the films I've seen, the poetry I've been taught and say, what have you given me? I look at the news and hear what people say about the news and say, what have you given me?

So, is that it? Have I said everything?

I don't think so. I've left out someone. I've left out something.

Me and my mind. Because as all those things were giving and giving and giving, I've been thinking and talking and writing. It's as if some kind of mincer, grater, mixer, cooker has churned all these things over. That made me too.

And even that's not everything either.

Really?

Yes, because I didn't make my mincer, grater, mixer, cooker that churns things over. It's my mind doing the mincing, grating, mixing and cooking. But I didn't make my own mind! I helped make it. Yes. While all those people and things were giving stuff to me.

We are made by others while we make ourselves. We make ourselves while others make us.

240

IF A COW DIDN'T FART FOR A WHOLE YEAR AND THEN DID ONE BIG FART, WOULD IT FLY INTO SPACE?

Mary Roach

author

It's true that cows produce a LOT of gas. Mostly methane, made by bacteria when they break down the grass inside the cow's gigantic, trash-can-size rumen (the cow's main stomach compartment). But guess what? Rumen gas – like any stomach gas – isn't farted. When we have a fizzy drink or a beer, the gas from the carbonation is burped out, not farted. Farts are made way down in the intestine, and in cows, there's relatively little digesting going on in that part of the body.

Guess what else: Not only do cows not fart, they don't burp. They are no fun at all at sleepover parties. Cows and other ruminants have a nifty trick that allows them to simply exhale the methane. My cow-fart-and-burp expert, animal science professor Ed DePeters of the University of California at Davis, explained how it works.

When a cow, say, or an antelope is feeling bloated and needs to make some room in her rumen, she blasts out some methane.

But instead of belching it straight up from her stomach – which would be noisy and might give away the animal's hiding place to a predator – she can shift things around and reroute the gas down into her lungs and then quietly breathe it out. Very dainty.

But let's not let this get in our way. Let's collect a year's worth of her methane breath. One cow produces about 187 pounds of methane gas in a year. Methane, by the way, is highly combustible. Which means it burns easily. Perfect! We'll store all the methane in a pressurised tank and use it to power a strap-on rocket for our fearless astro-cow.

To see how high she'd fly, I consulted a genuine rocket scientist called Ray Arons. Ray tested engines for the Apollo Lunar Module, the spidery-looking spacecraft that carried the astronauts to and from the surface of the Moon and was, he says, designed on the back of a napkin in a diner in Long Island, New York.

For our spacebound cow, Ray recommended a dual-nozzle engine for stability ('to avoid cow tipping') and a super-light-weight, aerodynamic, hi-tech flying suit to reduce air resistance (and look really boss at the pre-launch press conference). Then he got to work with his rocket-scientist formulas.

Ray calculated that 187 pounds of methane would supply two thousand pounds of thrust for about thirty-three seconds. He estimated that this would launch an aerodynamically streamlined 1,500-pound cow to an altitude of about three miles. Space begins around twenty miles, so the answer to your question is technically 'No.' Ray was impressed anyway. 'This methane engine is hot!'

WHY IS THE SEA SALTY?

Mark Kurlansky

journalist

People have always wondered why the sea is salty. From where does the salt come? Does it just come from the earth under the sea? Then why is that earth so much saltier than the earth in river beds and lake beds?

The first clue to the mystery is that river and lake water is salty. We don't notice its saltiness because it is so much less salty than the ocean.

So the first reason the sea is salty is that all the slightly salty water of all of the rivers of the world flows into the ocean and leaves its salt there. So all of the salt from the Earth's crust that is caught in rain water concentrates in the ocean. And then there is added salt from the ocean's floor, just as there is from a river bed, but this is a much larger bed.

This might make you wonder why the sea isn't getting steadily saltier. The main reason is that although salt is constantly coming in, there is also a flow of water that's not salty entering the ocean from rain, the mouths of rivers and melted ice. This is why the ocean is noticeably less salty near the mouths of rivers or areas where ice melts. On the other hand, ocean water is more salty than normal in places that are far from river mouths and where heat increases the level of evaporation. Just as salt companies make sea salt by leaving pools of sea water to cook in the sun until the water evaporates, the sea is noticeably saltier in the hot, tropical zone north and south of the equator.

There are other things that make some parts of the oceans very salty. In the late twentieth century scientists learned that in numerous spots on the ocean floor, water seeps into the earth and heats up. The trapped sea water cooks down to a higher concentration of salt and then is released into the ocean. A similar thing happens when volcanoes erupt under the ocean. The heat of molten rock also cooks and concentrates sea water.

On this planet rain runs into rivers, rivers flow into seas, and seas evaporate to form the moisture for new rains, which refill the rivers that refill the sea.

The Dead Sea, on the border between Israel and Jordan, is ten times saltier than the ocean because it is heated by a very strong sun with temperatures as high as 43 degrees centigrade and because its only water supply, the Jordan River, does not provide enough fresh water to keep it from cooking down. Eventually the Dead Sea will become a dry salt bed. It is the rivers and the ice and the rain that keep the oceans from concentrating and cooking away like the Dead Sea.

Scientists seem fairly certain that the oceans have maintained the same level of saltiness for hundreds of millions of years. But a new debate is emerging. If climate change causes a massive melting of the polar ice caps, will this not make the oceans less salty and change their ecosystems? More on this next century.

WHAT IS THE INTERNET FOR?

Clay Shirky

*who teaches people about the internet at
New York University*

The internet is only for one thing: letting computers talk to each other. (This includes mobile phones, which are little computers you can put in your pocket.) Everything we do on the internet, like playing games, or sharing pictures, or talking to our friends, uses connected computers.

Having one good way of connecting computers can let people do all sorts of things. This is different from the ways we used to watch or listen or talk to each other. TV was good for showing video, but it wasn't good for letting people talk to each other, and it wasn't good for seeing videos from other countries. Old-fashioned phones were good for letting two people talk to each other, but those phones were useless for letting ten people play a game together, or helping people look up words. What's good about the internet is that it helps any computer do all of those things.

What's even better about the internet is that people invent new things you can do on it all the time. When I started using the internet, *Minecraft* didn't exist, *Club Penguin* didn't exist, Facebook didn't exist. Neither did YouTube or Wikipedia. Even the Web didn't exist. Everything on the internet in those days was words – no pictures or sound.

In the last twenty years, all those things got invented by people who wanted to make computers do new things, even the Web itself. A man named Tim Berners-Lee had an idea for making Web pages that were connected to each other by links, and he used the internet to make it work.

In the next twenty years, more incredible things we can do on the internet will be invented. Maybe you will think of something you want computers to do, and then you can invent something on the internet too.

HOW DID MICHELANGELO GET SO FAMOUS?

Sister Wendy Beckett

art expert

Some people are famous for a few years and some are famous all their lives and some are famous even after their lives and a few, very, very few, are famous forever. Michelangelo is one of these very, very few. He was famous in his own lifetime, he is famous in our lifetime, and he will be famous in your great-grandchildren's lifetime.

Well, your question is: how? Why is he so famous? He is famous and will always be famous because he made marvellous paintings, especially the stupendous stories that he painted on the ceiling of an important church in Rome called the Sistine Chapel.

He carved wonderful figures too. The best-known are his great and beautiful figures of David and Moses, two heroes from the Bible. Even more beautiful is his carving of the Virgin Mary holding her dead son Jesus and grieving. It is called the *Pietà* (pronounced 'Pee-ay-ta').

When people look at these works of art they feel awe and

wonder. Sometimes tears of joy come into their eyes, because it is such an astonishing thing to be in contact with so moving a sign of what we humans can make. There is so much badness all around us, and here is something absolutely good.

But when a little person looks at something so great, they may not feel this sense of wonder. Really *seeing* what Michelangelo has done is not like turning on a switch. We have to grow into what we are seeing. Some big people are never able to do this. Their bodies have grown big but they are still little people inside. Usually this is because nobody has ever told them about art and what it can mean for us.

You are lucky because just in reading this, you are starting to learn. You already know that it is worth looking at a great artist like Michelangelo, and looking again, and going on looking, until one day you 'see'. Believe me, the day you 'see' the Sistine Chapel or the *Pietà* or David will be one of the most memorable days of your life.

HOW DO YOU FALL IN LOVE?

Falling in love is different for everyone. So we invited answers from three people who've thought about it a lot: two novelists who've written love stories, and a scientist who studies what goes on in our brains.

Jeanette Winterson

author

You don't fall in love like you fall in a hole. You fall like falling through space. It's like you jump off your own private planet to visit someone else's planet. And when you get there it all looks different: the flowers, the animals, the colours people wear. It is a big surprise falling in love because you thought you had everything just right on your own planet, and that was true, in a way, but then somebody signalled to you across space and the only way you could visit was to take a giant jump. Away you go, falling into someone else's orbit and after a while you might decide to pull your two planets together and call it home. And you can bring your dog. Or your cat. Your goldfish, hamster, collection of stones, all your odd socks. (The ones you lost, including the holes, are on the new planet you found.)

251

And you can bring your friends to visit. And read your favourite stories to each other. And the falling was really the big jump that you had to make to be with someone you don't want to be without. That's it.

PS You have to be brave.

David Nicholls

author

You can't make yourself fall in love, any more than you can decide to be taller or kiss your own elbow. Try it. You see? This can be a problem. Any number of broken hearts, sadness, disasters, wars even, might have been avoided if we were capable of controlling love.

Juliet could have ignored Romeo and learnt to love Paris. Henry VIII and Anne Boleyn might have made this really lovely couple. In one of my favourite books, called *Far from the Madding Crowd*, Bathsheba Everdene tells Gabriel Oak that she can't marry him because she doesn't love him, to which he replies: 'But I love you. And as for myself I'm content to be liked.' Which sounds reasonable enough. But being liked isn't the same thing at all. In the end, it just won't do. Anyone can be liked. The trick is to love and be loved in return.

So what's the difference between like and love? Sometimes I think of it as the difference between a cold and flu. Colds are common but flu is a much more serious business. Some people think they have the flu when really they've only got a cold. Some people know they've only got a cold, but exaggerate and try to pass it off as the flu.

I, for instance, was in a constant state of flu for a good twenty years. All I ever talked about was flu, flu, flu. Sometimes I was in flu with three or four different people at the same time. Looking back, I think I just had an awful lot of colds.

You may have noticed, round about that last sentence, that this comparison doesn't really hold up.

So while there's nothing you can do about falling in love, neither should you worry about it too much. Some things are going to happen whether you want them to or not. Your hair will go grey, your teeth will fall out, you will fall in love (though hopefully well before your teeth fall out). When it happens, don't panic. Stay calm. Try not to worry. Hope that they feel the same about you. If they do, then congratulations, you'll have a wonderful time for as long as it lasts. But if they don't love you back, then that's when the trouble really begins. Sorry.

Robin Dunbar

professor of evolutionary psychology

What happens when we fall in love is probably one of the most difficult things in the whole universe to explain. It's something we do without thinking. In fact, if we think about it too much, we usually end up doing it all wrong and get in a terrible muddle.

That's because when you fall in love, the right side of your brain gets very busy. The right side is the bit that seems to be especially important for our emotions. Language, on the other hand, gets done almost completely in the left side of the brain. And this is one reason why we find it so difficult to talk about our feelings and emotions: the language areas on the left side can't send messages to the emotional areas on the right side very well. So we get stuck for words, unable to describe our feelings.

But science does allow us to say a little bit about what happens when we fall in love. First of all, we know that love sets off really big changes in how we feel. We feel all light-headed and emotional. We can be happy and cry with happiness at the same time. Suddenly, some things don't matter any more and the only thing we are interested in is being close to the person we have fallen in love with.

These days we have scanner machines that let us watch a person's brain at work. Different parts of the brain light up on the screen, depending on what the brain is doing. When people are in love, the emotional bits of their brains are very active, lighting up. But other bits of the brain that are in charge of more sensible thinking are much less active than normal. So the bits that normally say 'Don't do that because it would be crazy!' are switched off, and the bits that say 'Oh, that would be lovely!' are switched on.

Why does this happen? One reason is that love releases

254

certain chemicals in our brains. One is called dopamine, and this gives us a feeling of excitement. Another is called oxytocin and seems to be responsible for the light-headedness and cosiness we feel when we are with the person we love. When these are released in large quantities, they go to parts of the brain that are especially responsive to them.

But all this doesn't explain why you fall in love with a particular person. And that is a bit of a mystery, since there seems to be no good reason for our choices. In fact, it seems to be just as easy to fall in love with someone after you've married them as before, which seems the wrong way round. And here's another odd thing. When we are in love, we can trick ourselves into thinking the other person is perfect. Of course, no one is really perfect. But the more perfect we find each other, the longer our love will last.

IF MY STOMACH WAS UNRAVELLED HOW LONG WOULD IT BE?

Dr Michael Mosley

science broadcaster

You can't unravel your stomach – it's a bag – but you can unravel your guts! Your guts, also known as your intestines, stretch all the way from your stomach right down to your bottom. That may not sound very far but the intestines are actually 8.5 metres long in an adult, a bit shorter in a child. If you had see-through skin you'd realise that they don't travel in a straight line but are coiled around themselves like a thin but enormously long snake.

Once you put food into your mouth you probably forget all about it, but that is just the start of a long and complicated journey. The first bit, which is the shortest, is called the oesophagus. It is about twenty-five centimetres long and is lined with extremely powerful muscles, which squeeze your food down into your stomach. These muscles are so powerful that you could eat standing on your head and the food would still keep going towards your stomach – though I'm not sure this is an experiment you would enjoy doing.

When the food reaches your stomach it is churned around and broken up, rather as if it were in a washing machine. The fluid in your stomach is about as acidic as a car battery. The acid is there to kill any bacteria that come in along with the food. The stomach itself is quite small. It's the size of a fist when it is empty but can expand to a medium-sized balloon.

After your food has been broken down by your stomach it is pushed out bit by bit into the small intestine, where it starts to get absorbed. The small intestine is about seven metres long; on average it is slightly longer in women than in men. It is covered in small hairs, called microvilli, which increase its surface area, allowing it to absorb more food. In fact, your small intestine alone has roughly the same surface area as a tennis court.

Once your food has got through the small intestine what's left of it passes into the large intestine. Here water is absorbed and bacteria get to work on the bits of food that have not already been broken down. The large intestine is much shorter, around 1.5 metres, and like the small intestine, it is lined with a mesh of cells called neurones, which are also found in your brain. Surprisingly, there as many brain cells in your guts as you'd find in the brain of a cat. You need them because, as we can see, digesting food is a complicated process.

After your intestines have absorbed everything useful from your meal, whatever's left over – as well as a lot of those bacteria – exits your body the next time you go to the loo. And that is the end of its long journey.

WHY DO WE HAVE AN ALPHABET?

John Man

author of books about writing

The first writing did not use an alphabet. The signs recorded whole words, like the icons on your computer. It can work pretty well. Suppose you want to record in pictures the famous line of Shakespeare, 'To be or not to be . . .', you might write:

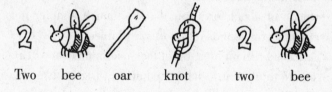

Two   bee   oar   knot   two   bee

Or you might write bits of words, which are called syllables. For example, 'belief', which you might spell in pictures as   —

bee leaf.

But it doesn't get you very far, because many words won't easily work in pictures. You can do it, but you need thousands of pictures because there are thousands of syllables! That's what happened with cuneiform, the triangle-shaped symbols they used in Mesopotamia ten thousand years ago, and later with Egyptian hieroglyphics and Chinese writing. It takes years for Chinese children to read very much. *You* could probably read by the time you were four or five. How did you do it?

The answer, of course, is that you learned the alphabet. That name simply means 'AB', the first two letters. And what is so special about the alphabet? It only uses twenty-six different signs, instead of thousands (well, fifty-two actually, because there are capital letters and small ones, but that doesn't bother us much). It is based on the fact that all the words you speak are made of about forty sounds. How much simpler it is to invent signs for sounds rather than syllables!

This is a brilliant idea, but it took over a thousand years for anyone to think of it. It happened about three thousand years ago, in Egypt, where those who could write were using hieroglyphs, writing with pictures and syllables. It happened that in Egypt there were tens of thousands of slaves who had been captured in today's Israel, Jordan, Syria and Lebanon. These foreigners, known as Asiatics, were probably mainly Hebrews, people we would now call Jews. Over the years, many had become much more than slaves. They were officials, craftsmen, and leaders in their own communities. They wished to write in their own language. Of course, they could have used hieroglyphs, but it was very difficult.

Someone, or perhaps a team, knew that hieroglyphics had a few signs that stood for what we would now call letters. There were about twenty-six of these signs. Each was a picture, but each also stood for its first sound. So they simply took the sign and translated what it meant.

One sign had something to do with an ox, so it looked like this: $\forall$. In 'Asiatic', an ox was *alep*, which began with an *a*. So back then a teacher might have said, '*a* is for *alep*,' as a teacher today might say, '*a* is for apple.' Over time, this ox sign changed to α, which became our little *a*.

Next, they took a sign which in Egyptian meant 'a shelter', and made it stand for *bayit* ('house'), and so on for about twenty-six letters. That's why we still have something like the *alep* or alpha and the *bayit* or beta as our first and second letters.

How do we know all this? Someone cut the signs into a rock, and other carvings nearby give clues to the date: about 1800 BC, some 3,800 years ago.

Once it was invented, the idea spread everywhere, through Greece to Europe, and through many other alphabets across the world. Today, the Chinese keep their own writing system, but they also use the alphabet, because it really is so much easier.

WHY DO I ALWAYS FIGHT WITH MY BROTHER AND SISTER?

Professor Tanya Byron

clinical psychologist

It is not unusual for siblings to fight – I used to fight with my sister when I was a child. We tend to fight with the people we are the closest to. Perhaps it's because we know that they will never stop loving us, however mean we are to each other.

Living together in the same space and being around each other all the time can also lead to disagreements over the way we do things and how we share in a way that maybe wouldn't happen with friends. However, fighting is not a good way to deal with upsets or disagreements. Instead of sorting a problem out it can make it worse as nasty things are said and done.

Also fighting siblings can cause stress and unhappiness for other family members and may lead to big rows with parents, which just makes everybody unhappy. Being part of a family helps us learn how to make relationships with people we love and also learn different skills important for life, including how to deal with anger and arguments.

You might find that your parents get angry with you for

fighting, and that might make you feel cross with them because they don't seem to understand why you are so upset. Actually your parents are probably upset with the *way* you are expressing your anger rather than the fact that you are angry. They might punish you for fighting because they don't want you and your siblings to think that this is the right way to try and sort problems out.

Name-calling and hitting are destructive ways to express anger. Little children often show frustration or anger by screaming or hitting because they have not learnt the words to say how they feel. However, as children get older we expect them to deal with disagreements by talking about the problem, so a solution can be found, rather than with nasty words or aggressive actions.

If you feel angry with your siblings try and take yourself away from the situation before you explode. Give yourself some time to calm down and then think about what is bothering you. You might find that once you're calm, the problem seems less of a big deal and so you can just ignore it and carry on being friends – some fights aren't worth fighting!

However, if you do feel very upset or hurt, try and tell them, and if you can't or they won't listen ask a parent or trusted adult to help you out. As I have got older I now see how my sister is the most important friend I have because we have grown up together and understand each other the best.

Remember, friends come and go but family are there for each other for life!

WHAT ARE RAINBOWS MADE OF?

Antony Woodward and Rob Penn

authors

Rainbows are made of light.

When sunlight shines through raindrops in the sky, the white light spreads out into bands of colour – red, orange, yellow, green, blue, indigo and violet. As the light enters the raindrop, it changes direction and is broken up. It's then reflected inside onto the back of the raindrop, and then broken up again into all the different colours when it leaves the raindrop.

For you to see a rainbow, it must be raining while the Sun is shining, and you have to be between the rain and the Sun. It's impossible to get to the end of a rainbow, which is a shame because, as we all know, pots of gold are buried there. It's impossible because although your eyes can see a rainbow, it is really nothing more than light shining through water droplets – it's not actually, physically there. Try walking towards a rainbow next time you see one: it just keeps moving further away.

Rainbows were only fully explained by Isaac Newton, a very clever scientist, three hundred or so years ago. Before that, for tens of thousands of years, people had the maddest ideas about rainbows. Some said they were paths connecting Earth to heaven; others thought a rainbow was the Sun God's belt, while a few thought the rainbow was an actual god appearing in the sky. One thing people have always agreed on is that rainbows are beautiful.

And how do you remember the colours of a rainbow? Red, orange, yellow, green, blue, indigo and violet. Try this: ROY G BIV.

WHEN DID PEOPLE START USING RECIPES?

Mario Batali

chef

People started using recipes the very first day they noticed that today's dinner was better than yesterday's. Before people were even able to write, they would have shown each other how to make their food taste better. If you think about it, even today with all the food shows on TV, cooking can still be learned by simply watching and repeating what you just saw, without checking the recipe on the internet, without a computer or even paper and pencil.

Roasting meat was probably the first of all recipes. It's likely that food as a cooked product started accidentally, when meat usually eaten raw fell into a fire that people were using to keep themselves safe and warm at night. We're not sure exactly when this happened but archaeologists have found ash and bone fragments from around one million years ago. Those people recognised that fire changed something in the food and remembered it for next time – almost certainly a kind of recipe, at least for successful eating.

One of the first real collections of recipes was made in the first century AD, by a food lover in the Roman Empire named Apicius. It was called *Re Coquinaria*, which translates loosely to 'About Food and Cooking' and was all about the magnificent cooking of the banquets of Ancient Rome.

The book was designed to be used in the kitchen by the servants and was organised very much like cookbooks now: divided by the main ingredients in the dish and the order that people ate them. Among the crazy dishes in style then were simple dishes like 'carrots with cumin sauce' or 'radishes with pepper', but also dishes like 'boiled whooping crane with herb and anchovy-honey sauce'. Of course many of these things are out of fashion today.

Nowadays there are cookbooks for every single type of cooking in the world and the magnificent local and regional cuisines from everywhere including the cooking of the Berber people in the Moroccan Atlas Mountains of the Sahara desert and even the pies of the members of the Rotary club in Topeka, Kansas.

WHY DOES THE MOON SHINE?

Dr Heather Couper

astrophysicist

The Moon is our companion in space. It's an amazing world, a quarter the size of the Earth. It is also very close, just 384,400 kilometres away. Remember, we're talking astronomy here! It only takes three days to get there by spaceprobe.

The Moon shines because it reflects light from our local star, the Sun. And because it circles the Earth once a month, we see different bits of it lit up as it follows its orbit. (The word 'month' actually comes from 'moonth'.) At 'New Moon', you can't see it, because it's directly in line with the Sun, and sunlight bathes the far side of the Moon. But as the Moon moves on, the Sun starts to catch its edge, lighting it up as a crescent.

If you have binoculars or even a small telescope, this is the best time to look at the Moon. The shadows are long and dark: they really emphasise the Moon's dramatic features. It is pockmarked with huge craters, caused by a massive bombardment by space rocks in the early days of the solar system. Because the Moon has hardly any atmosphere to wear its surface down, these scars are preserved in all their sharpness.

Full Moon is when the Moon shines brightest, opposite the Sun in the sky. Not the time to get out a telescope! But take a look at the 'face' of the 'Man in the Moon' without a telescope. You can see the Moon's 'eyes', 'nose' and 'mouth'. These are the dark, lava-filled scars that resulted from a hail of asteroids that hit the Moon 3.8 billion years ago.

Every year or so, something dramatic can happen. The Moon's orbit is angled to the Earth, and sometimes – at Full Moon – it can pass into the Earth's shadow. The brilliant Full Moon becomes 'eclipsed', or disappears. A truly spooky experience to see!

Amazingly, the Moon could shine a lot brighter in our skies if it was lighter in colour, and therefore more reflective. The Apollo astronauts who visited the Moon in the late 1960s and early 1970s were surprised at how dark the rocks were.

Neil Armstrong, the first man to set foot on the Moon, wrote that its colour was 'flicts'. If you think that's a made-up colour, you'd be right! An author that Neil Armstrong liked invented the word to describe a dark, sludgy brown that didn't have a name. And Neil thought 'flicts' fitted the Moon perfectly.

WHERE DO THE OCEANS COME FROM?

Dr Gabrielle Walker

writer and broadcaster on climate and energy

If you look at the Earth from space it's a beautiful blue planet and most of this blue colour comes from the oceans. There is actually more ocean than there is land on Earth, which makes ours a very watery world.

So where did all this water come from? Scientists don't know for sure, but they suspect that some may have come from the inside of the planet, and some from the outside.

Before any of the planets or even the Sun was born, 4,500 million years ago, we all started off as a swirling cloud of gas and dust, which also contained water. Eventually all this cosmic stuff started to stick together in clumps. The bigger the clumps got, the more strongly they tugged on the clumps around them, and we ended up with a whole solar system of planets – and the Sun in the middle.

But there were also quite big clumps left over – like a sort of builders' rubble – and these began slamming and crashing into the newborn planets like a giant pinball machine, making massive craters like the ones we see on the Moon, and heating up the Earth's surface so much that any water probably boiled off.

What happened next is that, over the years, comets also began smashing into the Earth. Comets are big dirty snowballs made almost entirely of ice, so when they landed they handed over their ice to start making oceans.

More water might then have been puffed out by volcanoes from the interior of the Earth where it was trapped inside rocks. Put all this together over millions of years and hey presto – you have your oceans.

By the way, the reason the water has somewhere to slosh around in is an entirely different story. The oceans sit in gigantic basins like sunken baths, which are much lower than the surrounding continents. That's because the continents, such as Europe, Asia and America, actually move around the surface of the Earth very slowly – about the rate your fingernails grow.

When two continents move apart they stretch out the space between them, which turns into a wide ocean basin like the Atlantic or Pacific, ready to receive all that water. When two continents move together they squeeze the space between them, until in some cases there is no ocean left. The mighty Himalayas were formed when two pieces of continent moved closer and closer until they swallowed up the ocean between them and then crashed together to make Mount Everest and all the other high mountains around it.

WHY DO SNAILS HAVE SHELLS BUT SLUGS DON'T?

Nick Baker

naturalist and broadcaster

Well, actually some slugs do have shells. Among them are some of the predatory slugs that we rarely see because they spend most of their lives underground chasing earthworms. Their shell is reduced to a tiny scale-like plate that sits on top of their body like a ridiculously small hat! So the line between slugs and snails is actually quite a fuzzy one.

Both snails and slugs are molluscs that belong to the gastropod family. But snails have evolved a handy portable capsule on their backs. This provides some form of protection from smaller predators, but more importantly, it allows them to slither into a world that is a little drier. The shell protects the delicate, moist body of the snail from drying out in the sun and the wind and it can be sealed up with a very tough layer of slime that dries out to form a front door called an epiphragm.

Slugs are much more vulnerable to becoming dehydrated as they don't have a shell. But what they are able to do is to go places where a shell would get in the way. They can squeeze

273

themselves into tight spots: cracks and crevices and even tunnels underground, places a snail wouldn't fit!

So slugs and snails have followed their own paths to come up with different solutions to the same problems of predators and exposure to the weather. They have found themselves different ways to make a living, something biologists call an ecological niche.

OUT-TAKES!

Comedians Stephen Fry, Sandi Toksvig, Clive Anderson, Robert Webb, Shazia Mirza, Sarah Millican and Jack Whitehall take on the experts with some alternative joke answers to the children's questions.

Stephen Fry

Why do elephants need trunks?
Elephants are very shy and modest and would blush to swim naked, the crocs and hippos would point at them. Me too, I even wear trunks in the bath.

Sarah Millican

Is it OK to eat a worm?
Only if your mam isn't watching.

Where does wind come from?
Brussels sprouts. And it wasn't me.

Are we all related?
Is this you angling for a Christmas present?

How does my cat always find her way home?
Catnav.

Why do we sleep at night?
Wait till you go to university, you'll be able to do it through the day too.

Do aliens exist?
Yes. They impersonate little brothers and sisters. So be careful.

Why do lions roar?
It's a yawn, you must be quite boring.

How did Michelangelo get so famous?
He was the only Teenage Mutant Ninja Turtle who could paint really well.

Do monkeys and chickens have anything in common?
Yes. They both taste delicious with chips.

How did they build the pyramids?
Out of loads and loads of Toblerones.

What is the internet for?
Before the internet, you had to talk to people, put kittens on pianos and push monkeys off logs yourself.

Sandi Toksvig

How are dreams made?
With egg white.

Why do people have different-coloured skin?
To make colour television more interesting.

Where does 'good' come from?
A small factory in Taiwan.

Why do we go to the toilet?
To get away from the dinner table.

Why are planets round?
To make them difficult to wrap.

Why is it dark at night?
To give torch manufacturers a sense of purpose.

Why do monkeys eat bananas?
Because they are a-peeling.

Why does the moon shine?
Beeswax applied once a month.

Why do we cook food?
To make having a kitchen in the house look less foolish.

What do you have to do to get into the Olympic Games?
Stop reading.

How do you make electricity?
By running in very tight nylon shorts.

Why do lions roar?
Anger management issues.

Why is the sun so hot?
To give sunblock manufacturers a sense of purpose.

What is gravity?
A drag.

Robert Webb

Who had the first ever-pet?
The Roman Emperor Julius Caesar. It was a squirrel called
Bianca.

Why do we have money?
We used to use cheese instead but it all got a bit messy.

Why is it dark at night?
Because the sun is charging.

Why does time go slow when you want it to go fast?
So you can pick your nose.

Are there any undiscovered animals?
Yes, the Belliphant, half bat, half elephant, will be discovered next Christmas in Hong Kong.

What are atoms?
Very very small peas made of science.

Why can't I tickle myself?
Because you would look mad.

How do cars work?
Small pigs run as fast as they can inside the wheels.

Can a bee sting a bee?
Yes, but then they have to go to bee prison where they have to dress up as flies, which they hate.

Why does the moon shine?
To try and get people to write songs about it.

What are rainbows made of?
Love. Rainbows happen when two clouds fall in love.

Why do we cook food?
To stop saucepans from getting bored in the cupboard.

What's inside the world?
A large gravity machine operated by Prince Charles wearing goggles.

What do you have to do to get into the Olympic Games?
Go to the Prime Minister's house and give him fruitcake.

Who invented chocolate?
King Arthur. His famous Round Table was actually a gigantic chocolate button which his knights would nibble at when he wasn't looking.

Why do scientists look at germs?
Because germs are always putting on terrific musicals.

Will we ever be able to go back in time?
Not until yesterday.

What would I look like if I didn't have a skeleton?
Like a jelly with hair.

Why do lions roar?
They're actually trying to sing but they're rubbish at it.

How does fire get on fire?
By being very naughty indeed.

Why is the sun so hot?
God was trying to cook a Scotch egg and he left it in the microwave for too long.

Do numbers ever stop?
No but they sometimes have a KitKat to keep them going.

Who named all the cities?
Ants. Ants have been naming cities since before even the oldest ant can remember. The first city was called Ant City, but since then they've got a bit more imaginative.

Why is water wet?
To make bathtime less scratchy.

How did Michelangelo get so famous?
He started his own fan club and had T-shirts made.

How do you fall in love?
Stand next to someone and see if you feel dizzy.

Why are some people taller than others?
The others just aren't trying.

Shazia Mirza

How are dreams made?
On full power for three minutes in a microwave.

Why are planets round?
Because they eat too many cupcakes.

Why do men grow beards and not women?
Because men need beards to store some food for later.
Women just keep it in their handbags.

Why can't we live forever?
Because we are needed in the next world to tidy up.

Why does time go slow when you want it to go fast?
Because time can read your mind and likes to annoy you.

How do trees make the air that we breathe?
By farting.

Why do we listen to music?
So that we don't have to listen to our mums.

Will we ever be able to go back in time?
Only if Doctor Who agrees to take us.

Why do we have an alphabet?
So we can have alphabet spaghetti.

How does my brain control me?
By sending messages to your belly button.

What is global warming?
It's when everyone in the world wears their favourite jumper.

Who named all the cities?
Man City.

Why is space so sparkly?
Because it eats glitter for breakfast.

Why can't animals talk?
Because they are always busy.

Why did dinosaurs go extinct?
Because they ran out of Pringles to eat.

If the universe started from nothing how did it become something?
By working hard.

Why do penguins live at the South Pole and not the North?
Because the South has the best hotels.

Why are some people taller than others?
Because they have secret ladders in their shoes.

Why is the sea salty?
Because of all the fish and chips.

Jack Whitehall

How are dreams made?
By the tooth fairy, to keep you distracted.

Is it OK to eat a worm?
Only if you eat it with red wine. White wine is for fish only.

Who had the first-ever pet?
Noah. He had a whole ark full of them and he lived till
he was nine hundred years old, working every day to earn
enough money to pay for his vet bills.

Why can't we live forever?
Because making one set of friends is difficult enough.

Clive Anderson

How are dreams made?
I don't know, I'll sleep on it.

Is it OK to eat a worm?
Not for the worm.

Who had the first-ever pet?
Eve. (A snake, not a success.)

Why do we go to the toilet?
To keep the rest of the house clean.

Why do men grow beards and not women?
Good question, I will try to grow some women immediately.

Why does time go slow when you want it to go fast?
It doesn't: it goes fast when you want it to go slow.

Why do monkeys eat bananas?
Because they aren't nuts.

Are there any undiscovered animals?
Yes, I think so but you can only prove it by then discovering them.

Why are the grown-ups in charge?
They got here first.

Why do we have an alphabet?
To make everything as easy as ABC.

How does lightning happen?
Quickly.

Who first made metal things?
Tintin.

Why are some people taller than others?
Because others are shorter than some.

CONTRIBUTORS

Maggie Aderin-Pocock has dreamt of space since she was a child, she now works as a space scientist and likes to tell people about the amazing universe we live in. Maggie encourages children to aim high by visiting schools to talk about her interesting job.

Jim Al-Khalili is a British scientist, author and broadcaster. He is a professor of physics at the University of Surrey and enjoys helping people understand science.

Benedict Allen has narrowly escaped death six times and lived and filmed in some of the world's most remote places. He has written about Amazon jungle adventures in *Mad White Giant* and New Guinea rituals in *Into the Crocodile's Nest*, and documented many other strange travel experiences.

Clive Anderson is a barrister turned comedy writer and presenter who first rose to fame hosting *Whose Line Is It Anyway?* and went on to front many radio and TV shows. He currently hosts *Loose Ends* and *Unreliable Evidence* on Radio 4, among others, and is the author of two *Unreliable Memoirs*.

David Attenborough is Britain's best-known natural history film-maker and environmentalist. His career as a naturalist and broadcaster has spanned nearly five decades and there are very few places on the globe that he hasn't visited.

Julian Baggini is the author of several books, most recently *The Ego Trick*. He is editor and co-founder of *The Philosophers' Magazine* and has written for numerous newspapers and magazines. Julian has also appeared as a character in two novels by Alexander McCall Smith.

Nick Baker collected jam jars full of spiders, ladybirds and frogs as a boy. Nowadays he is the 'bug man' who makes programmes about the much misunderstood world of creepy-crawlies. He is author of *The Bug Book*, for kids who love them too.

Mario Batali is the chef-owner of twenty restaurants around the world. He is the author of nine cookbooks and has appeared on television for the last fifteen years, most recently co-starring on *The Chew* airing daily on ABC. In 2008, Mario started the Mario Batali Foundation with the mission of feeding, educating and empowering children.

Wendy Beckett first appeared on TV as an art expert over twenty years ago. Viewers loved her enthusiasm and she was soon making major art series and publishing books. A nun from the age of sixteen, she went to Oxford University. She lives a life of quiet contemplation in the grounds of a Carmelite monastery.

David Bellamy, the botanist, writer and broadcaster, became one of Britain's best-loved TV personalities through his nature programmes of the 1970s and 80s. He has published thirty-four books, many for children, and founded the Conservation Foundation where he is still a director.

Vanessa Berlowitz has been making wildlife films for over twenty years with the BBC, including *Frozen Planet*, *Planet Earth* and *The Life of Mammals*. She has been lucky enough to film amazing things, from tiny spiders that hunt with James Bond-like intelligence to snow leopards in the mountains of Pakistan.

Heston Blumenthal is the chef who invented snail porridge and bacon-and-egg ice cream. He taught himself to cook, and likes experimenting with unusual flavours and techniques. A bit like Willy Wonka.

Liz Bonnin trained as a biochemist and wild animal biologist. She's often on TV presenting *Bang Goes the Theory*, and she recently fronted BBC1's *Super Smart Animals*. Liz loves big cats and works to help save tigers from extinction.

Alain de Botton writes books about philosophy, religion, art and travel. He's got a strange name because he was born in

Switzerland, but by now he speaks English fairly well. He is crazy about Lego and spends all his free time on the floor building stuff with his two sons Samuel and Saul, aged seven and five.

Derren Brown is a performer who combines magic and psychology to seemingly predict and control human behaviour, as well as achieving mind-bending feats of mentalism. He also writes books, paints portraits and has a deep love for parrots.

Dr Bunhead (aka Tom Pringle) performs spectacular live science demonstrations all over the world and on TV programmes like *Brainiac*. He also runs training courses for teachers on how to make science engaging as well as understandable.

Tanya Byron is a clinical psychologist, which means she helps treat people with mental health and behavioural problems. She regularly appears on TV and radio and writes many columns and books.

Mark Carwardine is a zoologist and outspoken conservationist. Author of more than fifty books, a wildlife photographer and magazine columnist, he presented BBC2's *Last Chance to See* with Stephen Fry and *The Museum of Life*, a series about the Natural History Museum in London.

Noam Chomsky is a linguist and philosopher based at the Massachusetts Institute of Technology, in the USA.

Marcus Chown writes books for grown-ups on things like black holes and the Big Bang, and very silly books for children, such as *Felicity Frobisher and the Three-Headed Aldebaran Dust Devil*.

Jarvis Cocker was frontman for the band Pulp for twenty-four years, becoming one of Britain's most cherished figures and bringing a rare, bookish wit to the pop charts. Now a solo artist, he writes songs, performs and has a radio show.

Heather Couper is a broadcaster and author on astronomy and space. She ran the Greenwich Planetarium for five years and has written over thirty books. In 1986, she was the astronomer onboard Concorde when it flew from London to New Zealand, showing the passengers Halley's Comet. Asteroid number 3922 is named 'Heather' in her honour.

Alex Crawford has been interrogated by intelligence agencies, rescued by the US army, and shot at on live TV. She is a special correspondent for Sky News, and author of *Colonel Gaddafi's Hat*, about reporting on war in Libya. Alex lives in Johannesburg with her husband Richard, one son and three daughters.

David Crystal has been described as a mixture of Gandalf and Dumbledore, thanks to his big white beard. In fact he is a professor who writes books and gives talks about the languages of the world.

Richard Dawkins is an evolutionary biologist who champions the teaching of evolution in schools. His many books include *The God Delusion* and *The Magic of Reality*, a science book for young people that explains the wonders of the universe in a way that's easy to understand.

Robin Dunbar leads a group of researchers studying the evolution of behaviour in monkeys, apes and humans.

David Eagleman is a brain scientist and writer. His brain laboratory studies time, the senses and the legal system.

Tracey Emin became famous as a 'Young British Artist' in the 1990s. A lot of her work tells stories about her own life. She draws and paints but has also sewn quilts and made art out of tents, beds, clothes and many other things.

Jessica Ennis is one of Britain's star athletes, specialising in hurdles and competitions where you have to be good at lots of different events. She is the current European and former world heptathlon champion. (Heptathlons have seven events including high jump, long jump, shot put and javelin.)

Mark Forsyth is a writer and journalist who blogs as The Inky Fool. The blog led to his quirky book *The Etymologicon*, an exploration of the hidden connections between words.

292

Russell G. Foster studies circadian neuroscience: the effects of night and day on humans and other animals. Russell knows all about your body clock: things like why you might be grumpy after bedtime and why teenagers get up so late.

Alys Fowler loves gardening and so does her dog, Isobel. (Though sometimes Isobel digs holes in the wrong place and Alys finds dog bones among the parsnips.) Trained in horticulture, she appears on gardening shows and writes books and columns.

Stephen Fry is an actor, writer and presenter. He is frequently referred to as a national treasure and modern-day Renaissance man.

A. C. Grayling is Master of the New College of the Humanities in London. He has written and edited over twenty books on philosophy and other subjects. Anthony ran away from school when he was fourteen, in protest at being caned too much, and is very glad that the cane is not used any more.

Lucie Green researches the atmosphere of the Sun, our nearest star. She writes about her discoveries, appears on TV and sometimes visits schools to talk with children who enjoy thinking about space too.

Susan Greenfield is a scientist who knows all about how brains work. She is particularly interested in what happens to your brain when you spend a lot of time playing computer games and using Twitter and Facebook.

Philippa Gregory was known as a historian when she wrote her first novel *The Other Boleyn Girl*, which was then made into a film. Six novels later, she is looking at the family before the Tudors: the magnificent Plantagenets. Her charity Gardens for The Gambia raises funds to provide wells for primary schools in Africa.

John Gribbin trained as an astrophysicist at Cambridge before becoming a full-time science writer. He has written scores of non-fiction books, among them *Time Travel for Beginners* and *In Search of Schrödinger's Cat*, as well as science fiction. John also writes songs for the band Three Bonzos and a Piano.

Bear Grylls has eaten maggots and slept in a deer carcass for his TV series *Man vs Wild* and *Born Survivor*. Trained in martial arts as a boy, he went on to join the British Special Forces and climbed Everest at just twenty-three. Bear has led many expeditions for charity in faraway places from Antarctica to the Arctic, and is Chief Scout in the UK.

Celia Haddon is the author of *Cats Behaving Badly* and *One Hundred Ways for a Cat to Train Its Human*. She recently wrote the biography of her current pet, *Tilly the Ugliest Cat in the Shelter*.

Claudia Hammond is a broadcaster, writer and psychology lecturer. She's written two books: *Time Warped* and *Emotional Rollercoaster* – a journey through the science of feelings. Claudia presents *All in the Mind* and *Mind Changers* on BBC Radio 4 and *Health Check* on BBC World Service.

Joanne Harris used to be a French teacher but now writes novels, including *Chocolat* and *The Lollipop Shoes*. She has also published two cookbooks. Joanne loves snorkelling, chilli and spaghetti Westerns but hates dancing.

Miranda Hart is a comedy writer and actress whose sitcom *Miranda* has made her one of Britain's favourite comedians. She wanted to be in comedy since she could remember but is not giving up on her other dream to be Wimbledon Ladies' Champion.

Adam Hart-Davis is a writer and broadcaster, former TV presenter of *Local Heroes*, *Tomorrow's World*, *What the Romans* (and others) *Did for Us* and *How London was Built*. He has read several books, and written about thirty. He enjoys woodwork and spends a lot of time making chairs, egg cups and spoons.

Roger Highfield is a director of the Science Museum Group, and before that he was the editor of *New Scientist* magazine. He's also known as the first person to bounce a neutron off a soap bubble.

Harry Hill used to be a doctor but not for a long time now. He's had many TV shows and tells jokes for a living. His hobbies are painting and drawing and the occasional game of swing-ball.

Richard Holloway went to a boarding school for trainee priests when he was just fourteen and went on to become Bishop of Edinburgh. He now makes programmes and writes books, most recently his own life story, *Leaving Alexandria*.

Kelly Holmes started running when she was twelve, encouraged by her school PE teacher. She began to dream of Olympic success and eventually won two gold medals at the 2004 Olympics, for the 800 and 1,500 metres. She encourages young people to fulfil their potential in sport or in life through her company Double Gold Enterprises and charity the Dame Kelly Holmes Legacy Trust. She was made Dame by the Queen in 2005.

John R. 'Jack' Horner is Curator of Palaeontology at the Museum of the Rockies, in the USA. He discovered the world's first dinosaur embryos and has two dinosaur species named after him. Jack was technical adviser for Steven Spielberg on all *Jurassic Park* movies, and on the Fox show *Terra Nova*. His dog is called Dawg.

Bettany Hughes is a historian who's fascinated by very old civilisations, particularly Ancient Greece. She's written books on *Helen of Troy* and the great thinker Socrates (*The Hemlock*

Cup). Her latest BBC TV series, *Divine Women*, is about amazing goddesses and warrior empresses.

Kate Humble is a TV presenter for wildlife and science programmes. She learnt to ride at the age of five and spent most of her early years mucking out horses. When not filming lions in Africa or lambing in Wales, she runs courses on countryside skills with her husband on their farm.

Simon Ings is a novelist, science writer and editor of *Arc*, a magazine about the future from the makers of the *New Scientist*. One of his books, *The Eye: A Natural History*, explored the chemistry, physics and biology of the eye.

Karen James is a biologist at Mount Desert Island Biological Laboratory in Bar Harbor in Maine in the USA. She is co-founder and director of the HMS *Beagle* Project, which aims to rebuild and sail the ship that carried Charles Darwin around the world in the 1830s.

Oliver James was the son of two psychologists and became one too. He makes programmes and writes articles and books, including *Affluenza* and *Britain on the Couch*. As a child he was pretty naughty and did terribly at school but still made it to university where he started working hard.

Sarah Jarvis is a GP (doctor) who regularly gives medical advice on BBC Radio 2 and *The One Show*, and writes newspaper and magazine columns. Her speciality is women's health.

Christian Jessen is a doctor, health campaigner and charismatic presenter of *Embarrassing Bodies*, *Supersize vs Superskinny* and *The Ugly Face of Beauty*. He works on Harley Street, the London street that is famous for medical practices. Christian also plays the oboe and sometimes gives concerts.

Annabel Karmel is an expert on what to feed babies and children, and mother of three herself. She wrote her popular *Complete Baby & Toddler Meal Planner* twenty-one years ago and has published twenty-five more books since, as well as presenting *Annabel's Kitchen* on TV.

Lawrence M. Krauss is a theoretical physicist at Arizona State University investigating big questions about the universe. He writes books, some of which kids read, like *The Physics of Star Trek*, and some of which dogmatic people don't, like *A Universe from Nothing*. Originally from Canada, he's keen on fly-fishing and mountain biking.

Mark Kurlansky is the author of twenty-five books, both non-fiction and fiction. His best known are *Cod* and *Salt* and their versions for children, *The Cod's Tale* and *The Story of Salt*. He has a daughter, Talia, who reads all his writing and tries to keep him from getting too boring.

Steve Leonard made his name as a TV vet on *Vet School* and has gone on to present wildlife series such as *Steve Leonard's Extreme Animals*, *Animal Kingdom* and *Safari Vet School*. He still can't believe his luck at getting so close to amazing animals in the wild.

Martyn Lyons moved to Sydney from Britain more than thirty years ago and now teaches at the University of New South Wales and writes history books. In Sydney, Christmas is the hottest time of year. His favourite Christmas memory is seeing Santa arrive at the beach in a surfboat.

George McGavin has had a lifelong obsession with wildlife, particularly insects. A leading entomologist (insect expert) and zoologist, he has written numerous books and after a long university career, he now presents science and natural history programmes for the BBC. George has several insect species named after him, and hopes they survive him.

Sally Magnusson is a journalist who presents the news in Scotland. The news can get a bit serious, so she also writes fascinating books like *Life of Pee: The Story of How Urine Got Everywhere*. Her first children's book is *Horace and the Haggis Hunter*, which her husband illustrated and her children helped to write.

John Man has two passions as a writer. One is the history of how we learned to write and make books. The other is Mongolia, because not many people know about it, and also because it's very important – it was the home of Genghis Khan, the greatest conqueror of all time.

Joanne Manaster loves sharing the wonders of science with young people. A biologist and former model, she teaches at the University of Illinois, writes for *Scientific American*, blogs and vlogs – that's video blogging.

Gary Marcus is a professor of psychology and director of the NYU Center for Language and Music. His books about the origins and development of mind and brain include *The Birth of the Mind*, *Kluge* and *Guitar Zero*, described as 'Jimi Hendrix meets Oliver Sacks'.

Sarah Millican is a British stand-up and best known for her own television series *The Sarah Millican Television Programme*. Her *Chatterbox Live* DVD is the highest selling DVD for a female comedian and in 2011 she won the 'Queen of Comedy Award' at the British Comedy Awards.

Shazia Mirza is a comedian and writer. She has written columns for the *Guardian* and *New Statesman*, has toured internationally and at the Edinburgh Fringe. She has appeared on *CBS 60 Minutes*, NBC's *Last Comic Standing*, *The Now Show* (Radio 4), and *Have I got News for You* (BBC).

Colin Montgomerie the golfer is one of his generation's most endearing sports stars. 'Monty' has held centre stage in forty-one tournament victories around the world, taken part in eight titanic Ryder Cup matches as a player and captained the winning 2010 European Ryder Cup team.

Michael Mosley has made many documentaries about the human body and medicine. He trained as a doctor but left to produce and present science programmes for the BBC. The latest include *Inside the Human Body* and *Frontline Medicine*, featuring the work of army surgeons in Afghanistan.

Steve Mould has a Masters in physics from Oxford and has appeared on *Blue Peter* as their science expert. He runs the Festival of the Spoken Nerd, a night of science and comedy that transferred to the West End in 2012, and takes science to music festivals like Glastonbury with Guerrilla Science.

David Nicholls is a novelist and writer for film and television. His popular first novel was *Starter for Ten* and the love story *One Day* has been read by millions worldwide. Both novels have been made into films with scripts written by David too.

Neil Oliver is an archaeologist and historian. He has become a familiar face presenting BBC's *Coast* and other programmes, and his most recent book is *A History of Ancient Britain*. Neil is happiest digging holes or watching Indiana Jones films.

Lorraine Pascale was spotted by a model scout when she was sixteen and became the first British black model on the cover of American *Elle* magazine. Although she had a fabulous time modelling, she changed career to follow her passion for cooking. Lorraine is now well known through TV series and her cookbooks, *Baking Made Easy* and *Home Cooking Made Easy*.

Nicholas Patrick, a British-born NASA astronaut, has flown on two shuttle missions and performed three spacewalks from the International Space Station. How did he get this cool job? Nicholas studied engineering at Cambridge and the Massachusetts Institute of Technology, and designed jet engines and aircraft cockpits.

Rob Penn has ridden a bicycle most days of his adult life. In his twenties, he gave up his job and cycled around the world. Rob is a journalist, TV presenter and author. He wrote about weather in *The Wrong Kind of Snow*. His latest book is *It's All About the Bike: the Pursuit of Happiness on Two Wheels*.

Robert Peston broadcasts and writes books about how people, businesses and countries make money, and why some become richer and some poorer.

Justin Pollard is a historian who spends most of his time writing books, articles and TV shows such as *QI*. He has advised

302

on movies from *The Boy in the Striped Pyjamas* to *Pirates of the Caribbean* and has also written nine books, one with an exploding toilet in it.

Christopher Potter is the author of *You Are Here, A Portable History of the Universe* – about the life of the universe, from quarks to galaxy super-clusters, and from slime to *Homo sapiens*.

Gavin Pretor-Pinney founded the Cloud Appreciation Society and is author of *The Cloudspotter's Guide*, *The Cloud Collector's Handbook* and *The Wavewatcher's Companion*. When he was young, he enjoyed asking questions. Now he enjoys answering them.

Philip Pullman is author of the trilogy *His Dark Materials* and many other novels. When Philip was eight he discovered the wonders of comic books, a force that's still a big influence on his writings and drawings today.

Gordon Ramsay had a career in football but retrained as a chef after an injury. He now owns restaurants from LA to Doha and has won many Michelin stars for them. You may have seen him doing some straight talking on TV shows like *Kitchen Nightmares* and *Hell's Kitchen*.

Martin Rees is the Astronomer Royal. In the old days, it was the Astronomer Royal's job to run the Greenwich Observatory in London. He doesn't do this, but is a professor at Cambridge University. He was lucky to become an astronomer at a time when we're learning so much about planets, stars and galaxies.

Joy S. Gaylinn Reidenberg is a professor of anatomy at the Mount Sinai School of Medicine in New York City. She studies the bodies of humans and animals, and is the comparative anatomist for the TV show *Inside Nature's Giants* where she looks inside really big animals to see how their bodies work.

Christopher Riley is a writer, broadcaster and film-maker specialising in astronomy and space flight. He has floated weightless on Russian and European space agency flights, and is a veteran of two NASA astrobiology missions chasing meteor storms around the Earth.

Mary Roach writes for *National Geographic*, *New Scientist*, *Wired*, and the *New York Times*. Her books include *Packing for Mars*, full of weird stuff about space travel. She enjoys backpacking, Scrabble, mangoes, and that *Animal Planet* show about horrific animals such as the parasitic worm that attaches itself to fishes' eyeballs.

Alice Roberts has always been fascinated by human anatomy and evolution. She currently teaches this subject at the University of Birmingham but also enjoys taking science to a wider audience: giving talks, writing books and presenting television programmes. Recent series on TV include *The Incredible Human Journey* and *Origins of Us*.

David Rooney works at the Science Museum in London. He looks after a big collection of things to do with transport. This includes aeroplanes, cars, bicycles, trucks and buses, as well as loads of models.

Michael Rosen's funny poems and stories are adored by children all over the world. Your parents probably grew up on the famous Chocolate Cake Story in *Quick, Let's Get Out of Here* and perhaps you liked *We're Going on a Bear Hunt*. Michael was Britain's Children's Laureate in 2007.

Meg Rosoff writes stories enjoyed by kids, teens and grown-ups. Her first novel, *How I Live Now*, has made readers shiver, laugh and cry, sometimes all at once. Meg's latest book is called *There Is No Dog*. It's about Bob, a nineteen-year-old boy who accidentally becomes God.

Marcus du Sautoy is a professor of mathematics at the University of Oxford. His many programmes about maths include *The Code* and some with comedians Alan Davies and Dara O Briain. He has also worked with Lauren Child (creator of *Charlie and Lola*), providing puzzles and codes for her books about child spy Ruby Redfort.

Roz Savage holds four world records for ocean rowing, including the first woman to row three oceans: the Atlantic, Pacific and Indian. She campaigns on environmental issues as a United Nations Climate Hero and is an Athlete Ambassador for 350.org. She has recorded her adventures in *Rowing the Atlantic: Lessons Learned on the Open Ocean*.

Rupert Sheldrake is a biologist and author of several books, including *Dogs That Know When Their Owners Are Coming Home*. He kept pigeons when he was ten and has always been interested in how animals find their way home.

Clay Shirky teaches at New York University, where he helps people figure out how to use the internet. He also writes books, including *Here Comes Everybody*, about what happens when lots of people use the internet to work together. He lives in New York City, with his wife and two kids just about your age.

Seth Shostak developed an interest in aliens at the age of eight, when he first picked up a book about the solar system. These days he is Senior Astronomer at the SETI Institute in California, which stands for the Search for Extra-Terrestrial Intelligence.

Daniel Simmonds is a primate keeper at ZSL London Zoo. Primates is the group name for many species of monkey and ape, and Dan works with all sorts: big gorillas, tiny squirrel monkeys, playful gibbons and cheeky spider monkeys.

Simon Singh wanted to be a nuclear physicist when he was nine. He studied particle physics and worked at Cambridge University and CERN, but realised he was better at writing about science than doing science. His books include *Big Bang*, *The Code Book* and *Fermat's Last Theorem*.

Gary Smailes is a military historian and writer of books for children, including the *BattleBooks* choose-your-own-adventure series. His favourite flour is self-raising.

Tim Smit created the Eden Project in Cornwall, with help from his friends. Over five years they turned a muddy pit into vast and beautiful gardens. Now thousands of people visit every year to see incredible plants and learn about the environment.

Dan Snow makes programmes about history for the BBC. He also writes books and iPad apps. He lives with his family and a giant Great Dane called Otto in the New Forest. Dan loves history because it includes the most exciting things that have ever happened to anyone ever.

Francis Spufford is a writer of non-fiction. He's mainly interested in history and in what different times have felt like to live in, but his latest book, *Unapologetic*, is about what religion feels like. He has two daughters, and is married to a vicar.

Iain Stewart is professor of geoscience communication at

Plymouth University. He has presented several popular pro-grammes for the BBC: *Earth – The Power of the Planet*, *How Earth Made Us*, *Men of Rock* and *How to Grow a Planet*.

Michaela Strachan has been presenting children's programmes and wildlife programmes for twenty-five years. During that time she has hand-fed sharks, rescued bears, kissed hum-mingbirds, run with cheetahs, caught snakes, been knee deep in bat poo and had her hand up an elephant's bottom!

Chris Stringer works in the palaeontology department of the Natural History Museum, London. So he knows a lot about early humans and how we evolved. When he was ten years old his two favourite things to draw were planes and human skulls.

Rosie Swale-Pope was fifty-seven when she decided to run around the world. Having lost her husband to cancer, she felt the need to 'grab life' and raise funds for charity. Rosie is the only person to have both sailed and run around the globe.

Kathy Sykes is a physicist and university professor who pres-ents fun science programmes and helped create the Chelten-ham Science Festival. She knows how to make a microscope from saucepan lids and a piece of glass, and once worked as a magician's assistant in Florence.

Sandi Toksvig is a leading comedian, actor and writer in Britain best known for political satire. She presents *The News Quiz* on BBC Radio 4 and has written thirteen books including *Hitler's Canary* for children and *Girls are Best*, a history book for girls.

Claire Tomalin has written historical and literary biographies of, among others, Mary Wollstonecraft, Jane Austen, Samuel Pepys, Thomas Hardy and Charles Dickens.

Peter Toohey lives in Canada on the end of the vast prairies near the Rocky Mountains. He is the author of *Boredom: A Lively History* and a professor of classics at the University of Calgary, though as a child he was very keen to become a farmer.

Joyce Tyldesley has worked on archaeological digs in Britain, Europe and Egypt. She has found many broken pots, and lots of stone tools, but has never found a mummy. When she is not working in Egypt she teaches Egyptology online to students of all ages, all over the world.

Gabrielle Walker writes books and makes programmes about the way the world works. She has swum with piranhas in the Amazon and used a hammer to pull lava out of a live volcano in Hawaii. But her favourite place is Antarctica and she hopes it stays cold and icy for a long time to come.

Robert Webb is one half of the comedy double act Mitchell and Webb and co-stars in *That Mitchell and Webb Look* and *Peep*

Show. Other work includes starring in the Dickensian spoof *Bleak Old Shop of Stuff* and the feature film 'The Wedding Video' as well as many other television series, panel and radio shows and theatre in the West End.

Jack Whitehall quickly established his talent on the live comedy circuit after being nominated for an Edinburgh Comedy Award at the 2009 Fringe Festival. Best known for his role as JP in the Channel 4 comedy *Fresh Meat*, Jack's acting career has also exceled with C4 giving him his own series *Hit The Road Jack* as well as BBC3 commissioning his self-penned *Bad Education*.

Jacqueline Wilson wrote her first 'novel' when she was nine, filling countless Woolworths' exercise books as she grew up. She's now written more than a hundred published books. The character Tracy Beaker is one of Jacqueline's most enduring creations.

Jeanette Winterson was adopted and grew up in a house where reading wasn't encouraged (apart from the Bible). Luckily the house had no bathroom, so Jeanette could read other stories by flashlight in the outside toilet. She wrote her first novel, *Oranges Are Not the Only Fruit*, when she was twenty-three and has been writing for grown-ups and children ever since.

Yan Wong is an evolutionary biologist and presenter on BBC1's *Bang Goes the Theory*, where he explains complicated stuff in a way that's easy to understand. His passion is biology and he helped write *The Ancestor's Tale* by Richard Dawkins.

Michael Wood is a historian, writer and film and TV producer, known for acclaimed books and TV series such as *Conquistadors*, *The Story of India* and most recently *The Story of England*.

Katie Woodard works as a forensic scientist using DNA traces to solve crimes in Seattle in the USA. She teaches her two kids at home and has written a children's book, *My First Book about DNA*.

Antony Woodward is author of *The Wrong Kind of Snow* and *Propellerhead*, on flying. His latest book is *The Garden in the Clouds*, about making a garden on top of a Welsh mountain. For some reason all his books so far have something to do with clouds.

Carl Zimmer has written thirteen books about science. His favourite animals are parasites. There is a tapeworm in an Australian fish that is named after him: *Acanthobothrium zimmeri*.

INDEX

strength, 154; undiscovered, 1,
204
Bentley, Wilson 'Snowflake',
109–11
Berners-Lee, Tim, 250
Big Bang, 77–8
birds: dinosaurs and, 26, 99–100;
feathers, 99; homing pigeons,
193–4; migrating, 194
black holes, 229
blood colour, 9–10
blue sky, 119–20
body, human: bones, 189–90;
brain, 33–4, 47–8, 101–2,
239; height, 211–12; skeleton,
161–2; stomach, 257–8; what it's
made of, 145–6
bones, 161–2, 189–90
Bonfire Night, 140
books, 69–70, *see also* writing
boredom, 219–20
brain, human, 33–4, 47–8, 101–2,
239
Brutus, 216–17
bubbles, 117

Cai Lun, 69
cakes, 27–8
camels, 132
carbon dioxide: greenhouse gas,
36, 164; trees and plants, 75–6

cars, 45–6
carvings, 143–4
Cassius, 216–17
cats: finding their way home, 193;
language, 42; pets, 50
chefs, 103–4
chickens: making dinosaurs, 100;
monkeys and, 123–4
child workers, 91–2
chimpanzees: bored, 219–20; brain
size, 101–2; communication, 41
chocolate, 171
city names, 155–6
clay tablets, 126
climate change, 35–6, 81, 164,
247
clouds: lightning, 207–9; rain,
97–8
colour: blood, 9–10; paint, 143,
144; rainbows, 265–6; seeing,
119–20; skin, 79–80; sky,
119–20
Columbus, Christopher, 171
comets, 272
computers, 247–8
concentrating, 121–2
Confucius, 69
cooking: cakes, 27–8; food, 55–6;
recipes, 103–4, 267–8
copper, 116
Cortés, Hernán, 171